Mathtastic Level 2 Numbers to 20 Teaching Book

Tracy Ashbridge

MEd, Grad Cert, PG Dip, BEd (Hons)

www.Mathtastic.com.au

Mathtastic

Learning Number Step by Step

Tracy Ashbridge

First Edition

2022 Brisbane

© Copyright 2022 Mathtastic: Tracy Ashbridge. All rights reserved

ISBN 978-0-6455822-1-5

The photocopiable pages, number problems booklet and word problems are available at:
http://mathtastic.com.au/mathtastic-resources-level-2/
Password: Mathtastic2

Contents

How to use Mathtastic 6

Module 1 – adding and subtracting 0,1,2,3 7
    Ideas for teacher lesson (1 hour) 7

Module 2 – add from largest, subtract by counting back/subtraction by count on 10
    Ideas for teacher lesson (1 hour) 10

Module 3 – rainbow facts 13
    Ideas for teacher lesson (1 hour) 13

Module 4 – **add and subtract 10's** 16
    Ideas for teacher lesson (1 hour) 16

Module 5 – double and halve 18
    Ideas for teacher lesson (1 hour) 18

Module 6 – near doubles 21
    Ideas for teacher lesson (1 hour) 21

Module 7 – partition by place value 24
    Ideas for teacher lesson (1 hour) 24

Module 8 – add and subtract by compensation 26
    Ideas for teacher lesson (1 hour) 26

Level 2 29

Numbers to 20 Resources 29
    Resources – not included 30
    Lesson Plan 31
    Homework - 15 mins per day 33
    Number Flips 35
    Rainbow numbers 42
    Number Track 1-20 43
    Number Line 1-20 44
    Ten Frames 5 wise 45
    Ten frames – pair wise 49
    Blank ten frames 53
    Number flips – pairs to 10 54

100 square	55
Bottom up 100 square	56
Go Fish Cards	57
Fingers Subitizing cards	59
Count in 5 cards	63
Go Fish – Teen numbers	65
Memory game place value	67
Rekenrek Subitizing	68
Count in 2's cards	71
Doubles Bus	73
Doubles and Halves skittles game	74
Follow on doubles and halves game	76
Number flips – doubles and near doubles	79
Near Doubles go fish or memory game	83
Counting in 10's cards	88
Arrow cards	89
Number track 10-20	93
Tally mark subitizing	94
Pocket Game	97
Level 2 Numbers to 20	99
Workbook	99
Instructions	101
Setting out the student book	102
Coding the answers	104
Module 1	105
Draw and solve	105
Draw and Solve	105
Module 2	106
Draw and solve	106
Retrieval Practice	106
Module 3	107

    Draw and solve...107

    Retrieval Practice 107

Module 4 108

    Draw and solve...108

    Retrieval Practice 108

Module 5 109

    Draw and solve...109

    Retrieval Practice 109

Module 6 110

    Draw and solve...110

    Retrieval Practice 110

Module 7 111

    Draw and solve...111

    Retrieval Practice 111

Module 8 112

    Draw and solve...112

    Retrieval Practice 112

Level 2 113

Numbers to 20 113

Problem Solving book 113

    How to use these cards..115

    Recording Page 116

    Module 1 – Add or subtract 0,1,2,3..117

    Module 2 – Add from the largest number, subtract by counting back 124

    Module 3 – Rainbow facts, pairs of numbers to 20...131

    Module 4- Add and subtract 10 138

    Module 5- Doubles to 20 145

    Module 6 – Near doubles...152

    Module 7 – Add using place value 159

    Module 8 – Add and subtract by compensation strategies 160

How to use Mathtastic

Mathtastic can be used to teach 1:1 or small groups. The programs spiral through different levels of numbers, each level addressing the 8 number sense strategies:

1. Add 0,1,2, 3, Subtract 0,1,2, 3
2. Add from largest number by counting on, subtract by counting back
3. Rainbow facts
4. Adding tens, subtract tens
5. Doubles/ halving
6. Near doubles
7. Partitioning numbers by place value
8. Adding and subtracting by compensating, Bridge 10 with a 9, Bridge 10 with 7 or 8, Round and adjust

There are 8 sections to each lesson.

1. Thinking problems – these are designed to be open ended and challenge the student to think mathematically. This is an opportunity for mathematical discussions and exploration.
2. Subitizing (sub rhymes with cube) – this is the skills of recognising a set of objects without counting and is a key skill which is not always established in students with difficulties in maths.
3. Counting patterns and objects – students need to develop a sense of the number line. This is a skill that students with difficulties are often weak in.
4. Number sense – each session there is a different focus working through the 8 areas of number sense. These are explained and modelled before applying the number sense concept to problems.
5. Game – many students with math difficulties can get anxious about maths and practising the skills through games is a less threatening way to gain the repetition they need. The games have been chosen to specifically practice the skill in focus and also allow for reasoning skills.
6. Word problems – students need to apply their knowledge in problems. These are organised **by the 11 different ways of presenting addition and subtraction problems so students don't** just learn to solve for the final answer but can be flexible to work around the problem.
7. Number problems – each session there are number problems related to the focus area and spaced retrieval of focus areas.
8. More games

Each module can be used as a lesson or can be split over several lessons depending on the time you have available and the speed the student works through the number sense strategies. The games can be repeated easily, and many have options to extend them.

Module 1 – adding and subtracting 0,1,2,3

Ideas for teacher lesson (1 hour)

Thinking Problems	**Thinking task** Cuisenaire Rods – how many different ways can you make 10 with 2 rods, can you organise them to make a pattern? What do you notice and wonder?
Subitizing	**Subitizing** Number flips – How many can you see on side 1? Use that information to know how many on side 2. If there are 6 on one side and 1 more on the other that will be 7. Students need to compare the patterns and how they have changed.
Counting	**Counting – patterns and objects** Build a 0-20 rainbow of numbers. Use this to support counting forwards and backwards from 20. As the student becomes more confident, turn some cards over to challenge them. Look at the teen numbers and discuss the patterns they can see and hear – 1 ten and ? ones, listen for the "teen" from 13-19. Check the student can hear the difference between "teen" and "ty" e.g., nineteen and ninety.
Number Sense	**Number sense focus area explain, explicit teach and model Examples and nonexamples** 1. To explain this concept, use a number line on the ground (a number line on a tarp can work well, or chalked onto the concrete). Ask students to add and subtract 0/1/2/3 by taking single steps (or extend by moving straight to the number if they can see how far 1,2,3 spaces are). 2. Teach 6+1=1+6. This is an important concept to understand. Start by showing this visually with counters. Then complete tasks where they match the number problems that are the same - memory game match the same problem just presented in reverse. During the

	number problem task, if you know this, this relates to starting with the largest number when adding. 3. Teach students how to solve simple missing number problems. Teach this using a <u>number track</u>. E.g., 6+? =8. Place a counter on 6, how many more do you need to add to get 8, count on the number track. Use a number track here as this is more concrete than a number line. Each box represents a number. Encourage students to work forwards and backwards along the number track to solve these problems.
Games	**Game/ hands on activity** Draw your own race. This game supports students to make the connection between a <u>number track</u> and a <u>number line</u>. You will need a number line 1-20, a number track 1-20 and a dice +1, +2, +3. Take turns to roll the dice and move your counter along both the number line and number track. The winner is the first to fall off the end of the number track and number line.
Word Problems	**Word Problems** These are all set out as Bett lines – read one line at a time and bet what the question is asking. **Problem Solving Book Level 2** – support with concrete materials – counters and drawing pictures. Track which types of questions the students can and can't do.
Number Problems	**A: Number problems** – draw the answer as well as number – pentagon Model 3+1=4 and 3-1=2, show in the 5 ways • Model - counters • Words – 3 counters plus 1 counter makes 4 counters • Pictures - draw • Equations - 3+1=4 • Contexts – make a number story **B: Retrieval and interleaving practice tasks – Level 2 book**

	**Game/ hands on activity** Make the most equations game You will need two 1-6 dice, paper, and pencil. Take turns to roll the 2 dice. Make as many equations as you can with the 2 numbers you roll. E.g., you roll 6 and 2- you could make 6+2=8, 2+6=8, 6-2=4 – this would score 3 points. If you rolled 2 and 2 you could make 2+2=4 and 2-2=0 which would score 2 points. An extension of this would be to use a 10- and 6-sided dice.

Module 2 – add from largest, subtract by counting back/subtraction by count on

Ideas for teacher lesson (1 hour)

Thinking Problems	**Thinking task** Cuisenaire Rods – extension - how many different ways can you make 10 with 3 rods, can you organise them to make a pattern?
Subitizing	**Subitizing** Revise ten frames pairwise and 5 wise from Level 1.
Counting	**Counting – patterns and objects** Count to 20 forwards and backwards. Use number rainbow to support if required. Turn over the numbers they know and leave those they need support with. Be aware that 14,13,12 can be tricky when counting backwards.
Number Sense	**Number sense focus area explain and model** **Examples and nonexamples** 1. Use a number track and practice counting on and back from a number. For addition, pick a card from a 0-20 (use rainbow number cards) pile and then roll a dice to pick the second number (+1, +2, +3, -1, -2, -3). For addition chose which number to start counting on from – the aim is for students to count on from the largest number, this will usually be the card but may not be if the card is very low. For subtraction chose which number to count back from – here count back from the largest number, if the subtract won't work, quickly roll again, no need to explain at this

	stage. Can the students do this by visualising a number line and not using concrete or pictorial supports? Encourage them to try this strategy.
	2. Revise the concept that 6+1=1+6
	3. Teach adding 3 numbers. Which 2 numbers would you add first and why? Reinforce with concrete materials that the **order of addition doesn't matter here to look for** ways to make it easier rather than just adding right to left.
	4. Teach complimentary addition for subtraction (count on to solve subtraction). Counting forwards is usually easier for students than counting backwards. In simple terms teach students to compare 2 numbers and find the distance between the 2 numbers. So, if you compare 11 and 9 to can find the difference by counting back. You can show the student these numbers are really close on the number line. Explain that subtraction is actually finding the difference/ how far apart the 2 numbers are. This then leads to solving 11-9 by counting forward on a number track to see the difference is 2. Do both count back and count forwards to show they have the same answer.
Games	**Game/ hands on activity** Subtraction from 19 game. You will need a pack of playing cards with the cards 1-5 (remove cards 6-10 and the picture cards). Players take turns to pick a card from the pack. The player then subtracts the card they have picked from 19. If correct they keep the card. (Discuss the strategy they used or prompt a strategy if required, this will depend on the number drawn). Use concrete materials and/or number tracks to support development of a mental number line. Variation – use more cards and/or change the target number to another teen number. An extension to this game would be to include all the cards from 1-10.
Word Problems	**Word Problems** These are all set out as Bett lines – read one line at a time and bet what the question is asking. Problem Solving Book Level 2 – support with concrete materials – counters and drawing pictures. Track which types of questions the students can and can't do.

 Number Problems	A: **Number problems – draw the answer as well as number – pentagon** Model 3+1=4 and 3-1=2, show in the 5 ways - Model - counters - Words – 3 counters plus 1 counter makes 4 counters - Pictures - draw - Equations - 3+1=4 - Contexts – make a number story Teach vertical and horizontal presentation of equations. B: **Retrieval and interleaving practice tasks – Level 2 book**
 Games	**Game/ hands on activity** Snap it game – use unifix cubes, Lego will work as an alternative. Start with 10 cubes snapped together. One player takes the cubes and "snaps" them behind their back. They then **reveal** one part, and the partner has to work out how many cubes in the other part of the "snap." Alternatives, use a different number of cubes to "snap"

Module 3 – rainbow facts
Ideas for teacher lesson (1 hour)

	Thinking task Caterpillar count – imagine you have 10 caterpillars in a jar but each time you turn around a few crawl out. Explore different numbers crawling out and how many are left in. How many caterpillars and in the jar and how many are on the leaf? Investigate different ways the caterpillars can be arranged when some are in the jar and some out.
	Subitizing Review 10 frames pairs to 10. Flip flop charts with 10 pictures – if I can see 6, how many can you see? (4).
	Counting – patterns and objects Count to 50, listen to the patterns. Use 100 square to support identifying the pattern. What do you notice and wonder? Colour code what stays the same (red) and what changes (green). Compare both vertically and horizontally. 2 different 100 squares are included for this activity – top down and bottom up.

 Number Sense	**Number sense focus area explain and model** **Examples and nonexamples** Build 2 ten frames with 2 different coloured counters [ten frame diagrams] Use the ten-frame pattern to discuss which pairs go together and why. Here you can see 8+2 but also 18+2. Look for the pairs to 10 even past 10 e.g., 11+9 – you can see 1+9 there too. A good way to explain this is by using a 1-20 bead string (organised in alternating colours in 5's). Slide and hide a few from the end of the bead string and then discuss how many you can see and how do you know. Use the different colour sections to use subitizing to support counting how many you can see. Another way to explain this is using Cuisenaire rods: tens and a colour – what do you need to add to make 20. Relate this back to pairs to 10. Compliments ping pong – make a number 1-19 – the student replies with the complimentary number. Then swap roles.
 Games	**Game/ hands on activity** Go Fish - pairs to 20. Shuffle up the 2 sets of the number cards 0-20. Deal 7 cards per player and put the remaining cards in the centre. Before starting the game, the players check if they already have any pairs to 20 and put them aside. The first player asks another player if they have the matching pair to a card they have. E.g., if the player has 15, they would ask another player if they have 5. If the other player has 5, they hand it over. If not, they say, "go fish" and the player collects a card from the pile. The game ends when 1 player has no cards left.

Word Problems	**Word Problems** These are all set out as Bett lines – read one line at a time and bet what the question is asking. Problem Solving Book Level 2 – support with concrete materials – counters and drawing pictures. Track which types of questions the students can and can't do.
Number Problems	**A: Number problems – draw the answer as well as number – pentagon** Model 3+1=4 and 3-1=2, show in the 5 ways - Model - counters - Words – 3 counters plus 1 counter makes 4 counters - Pictures - draw - Equations - 3+1=4 - Contexts – make a number story **B: Retrieval and interleaving practice tasks – Level 2 book**
Games	**Game/ hands on activity** How many are hiding? You need 20 small items (cubes or marbles will work well) and a bowl (must not be see through). Player 1 hides some of the items under the bowl and leaves the rest visible on the table. Player 2 has to work out how many items are under the bowl and explain how they know.

Module 4 – add and subtract 10's
Ideas for teacher lesson (1 hour)

Thinking Problems	**Thinking task** Cuisenaire rods connecting +/-. Create number sentences about the colours to show the link between addition and subtraction. e.g., purple + yellow = blue blue-purple = yellow
Subitizing	**Subitizing** Show fingers and hands cards up to 20. How many can you see? Explain how you know.
Counting	**Counting – patterns and objects** Count in 5's using a 100 square to see the numbers. After a few practices forwards and backwards. Order the numbers on cards along the floor. Jump along forwards and backwards saying the pattern. Turn over any cards the student knows until there are only a few left.
Number Sense	**Number sense focus area explain and model** **Examples and nonexamples** Partition numbers into tens and ones. Start with a teen number in counters and show how to regroup as a group of ten and a group of ones. Explain we can show these tens and ones in many different ways. MAB, Rekenrek, Cuisenaire, arrow cards. Show the same number in all these different ways. Many students don't make the link from the concrete to the numbers.

Games	**Game/ hands on activity** Go Fish – teen numbers. Shuffle up the number cards for this game. Deal 7 cards per player and put the remaining cards in the centre. Before starting the game, the players check if they already have any pairs and put them aside (a pair here is the sum and the answer e.g., 13 would pair with 10+3). The first player asks another player if they have the matching pair to a card they have. E.g., if the player has 15, they would ask another player if they have 10 +5. If the other player has 10 + 5 they hand it over. If not, they say, "go fish" and the player collects a card from the pile. The game ends when 1 player has no cards left. You could also use these cards and play memory game.
Word Problems	**Word Problems** These are all set out as Bett lines – read one line at a time and bet what the question is asking. Problem Solving Book Level 2 – support with concrete materials – counters and drawing pictures. Track which types of questions the students can and can't do.
Number Problems	**A: Number problems – draw the answer as well as number – pentagon** Model 3+1=4 and 3-1=2, show in the 5 ways • Model - counters • Words – 3 counters plus 1 counter makes 4 counters • Pictures - draw • Equations - 3+1=4 • Contexts – make a number story Teach vertical and horizontal presentation of equations. **B: Retrieval and interleaving practice tasks – Level 2 book**
Games	**Game/ hands on activity** Memory game – match visual representation to number e.g., 14 in MAB would match 14 as a number. To speed up the game turn over 3 cards.

Module 5 – double and halve
Ideas for teacher lesson (1 hour)

Thinking Problems	**Thinking task** Happy Halving – how many different ways can you half a square? You can do the obvious ways, but can you think outside the box? It may help to use squares made on grid paper.
Subitizing	**Subitizing** Rekenrek – use the picture cards to subitize using the Rekenrek. How do you know? What patterns can you see?
Counting	**Counting – patterns and objects** Count to and from 20 in 2's. Use a number track first to establish the pattern. Extend by ordering number cards long the floor and jumping along the track saying the numbers forwards and then backwards. As the student learns the pattern turn the cards over.
Number Sense	**Number sense focus area explain and model** **Examples and nonexamples** 1. Use the double decker bus to model the concept of doubles.

	2. Use the function machine in both directions to teach double and half are opposites. Put in 4 and double it to make 8, in reverse, half 8 and it becomes 4
	3. Teach concept of half using Cuisenaire rods e.g., 2 yellow (5) = 1 orange (10).
	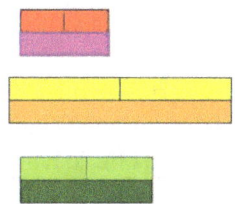
	4. This is also a good visual way to show half. You can also use grid paper to show half above he line and half below.
	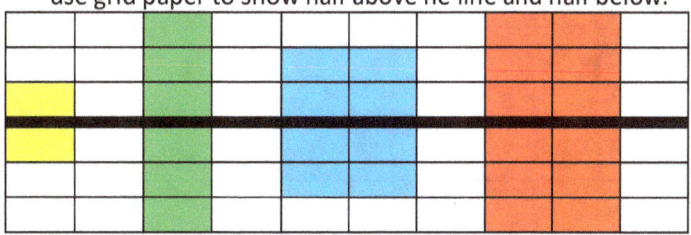
	Here you can see visually, 1+1=2, 3+3=6 etc. You could also make this with counters, just draw a line across the page.
	**Game/ hands on activity** Play the double and halves skittles game. First round, write numbers from 2,4,6,8,10,12,14,16,18,20 on the skittles and then double the number on the dice. For the halving version, write numbers 1-10 on the skittles but when you roll you have to say what number was halved to get that number e.g., if you roll 8, that is half of 16.

Word Problems	**Word Problems** These are all set out as Bett lines – read one line at a time and bet what the question is asking. Problem Solving Book Level 2 – support with concrete materials – counters and drawing pictures. Track which types of questions the students can and can't do.
Number Problems	**A: Number problems – draw the answer as well as number – pentagon** Model 3+1=4 and 3-1=2, show in the 5 ways • Model - counters • Words – 3 counters plus 1 counter makes 4 counters • Pictures - draw • Equations - 3+1=4 • Contexts – make a number story **B: Retrieval and interleaving practice tasks – Level 2 book**
Games	**Game/ hands on activity** Follow on game for doubles and halves.

Module 6 – near doubles
Ideas for teacher lesson (1 hour)

Thinking Problems	**Thinking task** 16 fences – use 16 same size lengths of paper or rods. How many different ways can you build a fence for Buster. Which is the biggest? Which is the smallest? Extend – try 24 or 36 pieces of fence.
Subitizing	**Subitizing** Number flips with doubles and near doubles on the back. For example, if the card shows 8 – the back would show 17 or 15 depending on if it is double plus 1 or double minus 1.
Counting	**Counting – patterns and objects** Count in 2's – odd numbers. Use the number track or 100 square to support if needed. At this stage we are expecting that students are developing an internal number line. If this is still insecure, use visual supports to ensure the internal number line they are creating is accurate. Before starting this activity check the student knows odd and even numbers. Build odd and even numbers on ten frames pairwise. Odd numbers have an odd counter and don't sit nicely in pairs. Even number Odd number

Number Sense	**Number sense focus area explain and model** **Examples and nonexamples** Use a variety of materials to show odd and even numbers visually. E.g., Cuisenaire, doubles bus, numicon, counters.
Games	**Game/ hands on activity** Go Fish- near doubles. Shuffle up the number cards for this game. Deal 7 cards per player and put the remaining cards in the centre. Before starting the game, the players check if they already have any pairs and put them aside (a pair here is the sum and the near doubles sum e.g., 13 would pair with 6+7). The first player asks another player if they have the matching pair to a card they have. E.g., if the player has 15, they would ask another player if they had 7 + 8. If the other player has the card, they hand it over. If not, they say, "go fish" and the player collects a card from the pile. The game ends when 1 player has no cards left.
Word Problems	**Word Problems** These are all set out as Bett lines – read one line at a time and bet what the question is asking. Problem Solving Book Level 2 – support with concrete materials – counters and drawing pictures. Track which types of questions the students can and can't do.
Number Problems	**A: Number problems – draw the answer as well as number – pentagon** Model 3+1=4 and 3-1=2, show in the 5 ways - Model - counters - Words – 3 counters plus 1 counter makes 4 counters - Pictures - draw - Equations - 3+1=4 - Contexts – make a number story **B: Retrieval and interleaving practice tasks – Level 2 book**

|  | **Game/ hands on activity**

Double take game.
You will need a pack of playing cards with cards 2-10 from each of the 4 suits.
Place 9 cards down in a 3x3 array. You can then play out any cards which are doubles or near doubles. You then replace your used cards ready for the next round.
With these cards:
4 doubled is 8
5 doubled is 10
8+9 is a near double
5+6 is near double
This just leaves me with 1 card (ace of diamonds) for the next round.
 |

Module 7 – partition by place value
Ideas for teacher lesson (1 hour)

Thinking Problems	**Thinking task** Sums investigation Put the cards 7/8/9/10 into bag. Pull out 2 cards and add them together. Replace the cards and pull out 2 more cards. After a few goes, stop, and consider how you will know when you have all the possible combinations to add together.
Subitizing	**Subitizing** Fingers and hands to 20 How many fingers are showing? How do you know? What patterns can you use?
Counting	**Counting – patterns and objects** Count in 10 to 100 forwards and backwards. Use 100 square to support learning the pattern. Extend by ordering number cards long the floor and jumping along the track saying the numbers forwards and then backwards. As the student learns the pattern turn the cards over.
Number Sense	**Number sense focus area explain and model** **Examples and nonexamples** Use arrow cards, MAB, numicon and Rekenrek to explore the place value of numbers 11-19. Review "teen" vs "ty" auditory discrimination. It is important that the student knows that 15 is made up of a 10 and 5.

Games	**Game/ hands on activity** Go Fish – Teen numbers Revisit this game as Go Fish and/ or memory game
Word Problems	**Word Problems – at this level these are the same as +/- 10 problems** These are all set out as Bett lines – read one line at a time and bet what the question is asking. Problem Solving Book Level 2 – support with concrete materials – counters and drawing pictures. For this module finish any outstanding problems. Track which types of questions the students can and can't do.
Number Problems	**A: Number problems – draw the answer as well as number – pentagon – at this level these are the same as +/- 10 problems** Model 3+1=4 and 3-1=2, show in the 5 ways • Model - counters • Words – 3 counters plus 1 counter makes 4 counters • Pictures - draw • Equations - 3+1=4 • Contexts – make a number story **B: Retrieval and interleaving practice tasks – Level 2 book**
Games	**Game/ hands on activity** **Marching on game** – You need number strip from 10-120 for each player marked in 10's and a spinner with numbers 1-5. You also need a pack of playing cards with the picture cards removed. Player 1 spins the spinner and picks up the number of cards shown. If any of those cards, make a pair to 10 then they can cover 10 more on their strip. The cards are returned to the bottom of the pile and player 2 has their go. If the cards allow you may cover more than 1 group of tens in a go. An extension of this game would be to play in reverse and to subtract groups of 10 OR use a different 12 tens numbers e.g., 60-180, 230-250.

Module 8 – add and subtract by compensation
Ideas for teacher lesson (1 hour)

	Thinking task Pocket game. You need a game board with numbers 1-20 and 2 sets of number cards 1-20 and 2 players. Place the pile of cards in the centre. Each player takes a card. The player with the biggest number takes both cards and places them on the game board for the biggest number. E.g., players pick 7 and 9 – the cards are both placed on 9 on the game board.
	Subitizing Tally marks to 20 – how many can you see? How do you know? What patterns can you see?
	Counting – patterns and objects During this book students have been developing their understanding of the mental number line. The next step is to be able to create their own number lines. 1. Provide the student with a blank number line and label 0 and 10. Can the student write in the numbers? Can they write them in with the correct spacings? 2. Extension – do the same with a blank number line 0-20 3. Further extension – do parts of the number line e.g., from 5-10 or 12-18.
	Number sense focus area explain and model **Examples and nonexamples** In this module, the students are developing their ability to use compensation to support their addition and subtraction. The following activities can be done with Rekenrek, bead strings, Cuisenaire (on a number track) and empty number lines. 1. Start with asking the students to work out 9+6 – they could do this by counting on from 9. However, in this

module we want them to use compensation strategies. Here we are looking to model that if you move 1 from the 6 to the 9 the problem becomes 10+5 which is easier to solve. There are lots more examples of questions to support this strategy in the Number Problems book for this module.

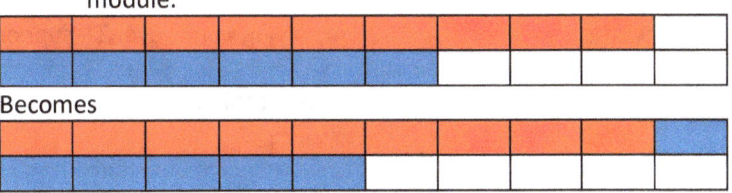

Becomes

2. Working on an empty number line students can show how they break down a number for addition e.g., 8+6. So here the student breaks down the 6 into 2 - to make to 10 and then 4 more to total 16.

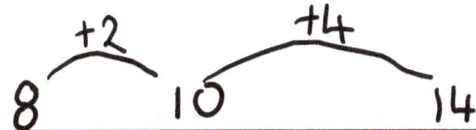

3. Bridging on a number line works well for students who are struggling with addition. This complimentary addition means they can work forwards along the number line.

e.g., 15-7. So here the student works forwards – we have already taught that the distance is the same regardless of working forwards or backwards along the number line. So, the count forwards here (3+5) equals 8 which is the same as if we had subtracted 7 from 15.

Game/ hands on activity

Minimise (or Maximise) subtraction game
You need playing cards 1-10 (remove the picture cards). Players take turns to pick up 4 cards and arrange them to minimise (or maximise) the difference between the 4 cards. E.g., if the player pulled out 1, 4, 6, and 8.
Minimise the difference 4-1=3, 8-6=2
Maximise the difference 8-4=4, 6-1=5

	Students record their answers to practice writing equations correctly and number orientation.
Word Problems	**Word Problems** These are all set out as Bett lines – read one line at a time and bet what the question is asking. Problem Solving Book Level 2 – support with concrete materials – counters and drawing pictures. For this module finish any outstanding problems. Track which types of questions the students can and can't do.
Number Problems	**A: Number problems – draw the answer as well as number – pentagon** Model 3+1=4 and 3-1=2, show in the 5 ways • Model - counters • Words – 3 counters plus 1 counter makes 4 counters • Pictures - draw • Equations - 3+1=4 • Contexts – make a number story **B: Retrieval and interleaving practice tasks – Level 2 book**
Games	**Game/ hands on activity** Roll 3 six-sided dice and add them together. Which numbers did you add together first and why? This leads to lots of discussion about strategy. Model the strategy on an empty number line. For subtraction, roll 3 six-sided dice and then subtract from 20 – which numbers did you subtract first and why? Model the strategy on an empty number line.

Level 2

Numbers to 20

Resources

The photocopiable pages, number problems booklet and word problems are available at:
http://mathtastic.com.au/mathtastic-resources-level-2/
Password: Mathtastic2

Resources – not included

Rekenrek 1-20

Dice 1-6

Dice 1-10

Blank 6-sided dice

Bead string 1-20 (in 2's and 5's)

Cuisenaire rods

Numicon – optional

Counters

Playing cards

Lesson Plan

Thinking Problems	**Thinking task**
Subitizing	**Subitizing**
Counting	**Counting – patterns and objects**
Number Sense	**Number sense focus area explain and model** **Examples and nonexamples**
Games	**Game/ hands on activity**

Word Problems	**Word Problems** These are all set out as Bett lines – read one line at a time and bet what the question is asking. Problem Solving Book Level 2 – support with concrete materials – counters and drawing pictures. Track which types of questions the students can and can't do.
Number Problems	**A: Number problems – draw the answer as well as number – pentagon** Model 3+1=4 and 3-1=2, show in the 5 ways • Model - counters • Words – 3 counters plus 1 counter makes 4 counters • Pictures - draw • Equations - 3+1=4 • Contexts – make a number story **B: Retrieval and interleaving practice tasks – Level 2 book**
Games	**Game/ hands on activity**

Homework - 15 mins per day

Day 1	
Counting	
Number Problems	
Games	
Day 2	
Counting	
Problem Solving	Word Problem solving booklet - Level 2
Games	

Day 3	
Counting	
Number Problems	Number problems booklet Level 2– draw the answer as well as the calculation
Games	
Day 4	
Counting	
Problem Solving	Word Problem solving booklet - Level 2
Games	

Number Flips
Cut across the page and then fold down the middle so you can open and compare if needed.

4 penguins	5 penguins
5 penguins	6 penguins
6 penguins	7 penguins
7 penguins	8 penguins

35

8	8
8	10
9	9
11	12
12	13

13	15
14	20
15	19
16	18
17	19

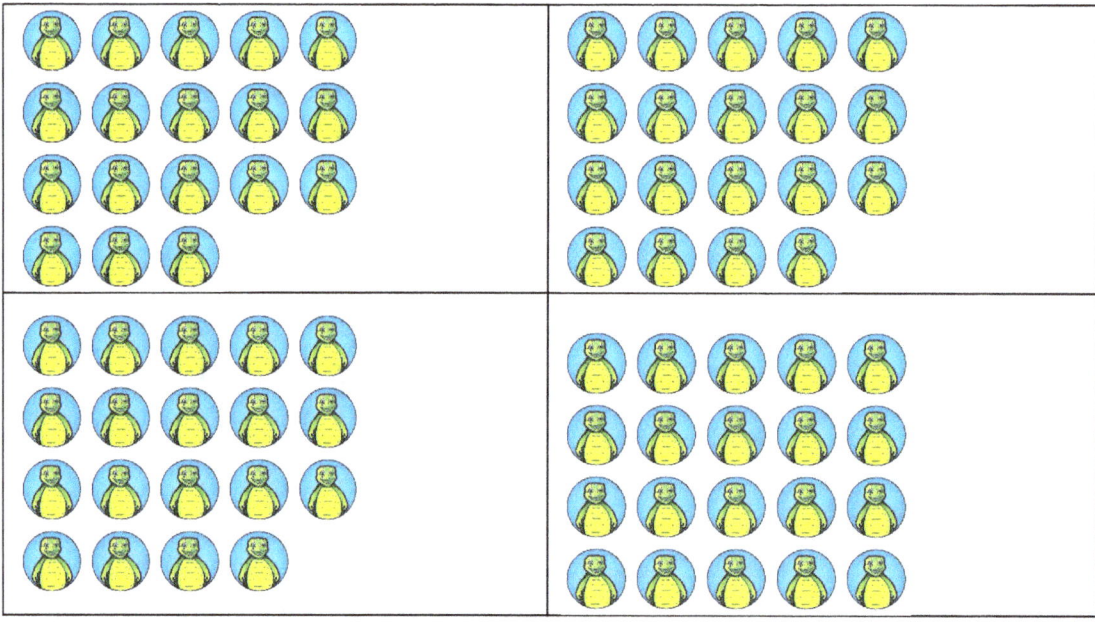

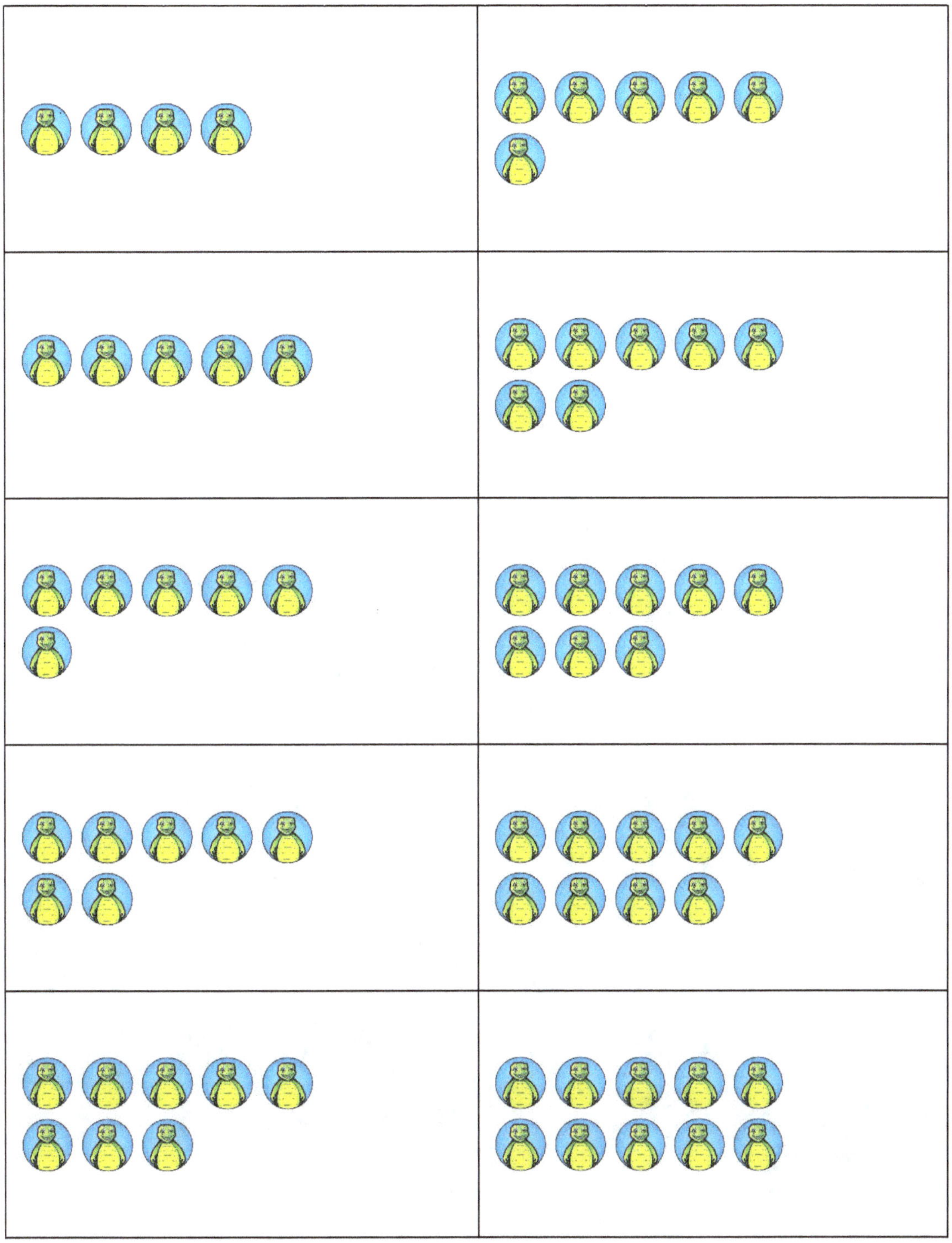

🐢🐢🐢🐢	🐢🐢🐢🐢🐢 🐢
🐢🐢🐢🐢🐢	🐢🐢🐢🐢🐢 🐢🐢
🐢🐢🐢🐢🐢 🐢	🐢🐢🐢🐢🐢 🐢🐢
🐢🐢🐢🐢🐢 🐢🐢	🐢🐢🐢🐢 🐢🐢🐢🐢
🐢🐢🐢🐢🐢 🐢🐢🐢	🐢🐢🐢🐢🐢 🐢🐢🐢🐢🐢

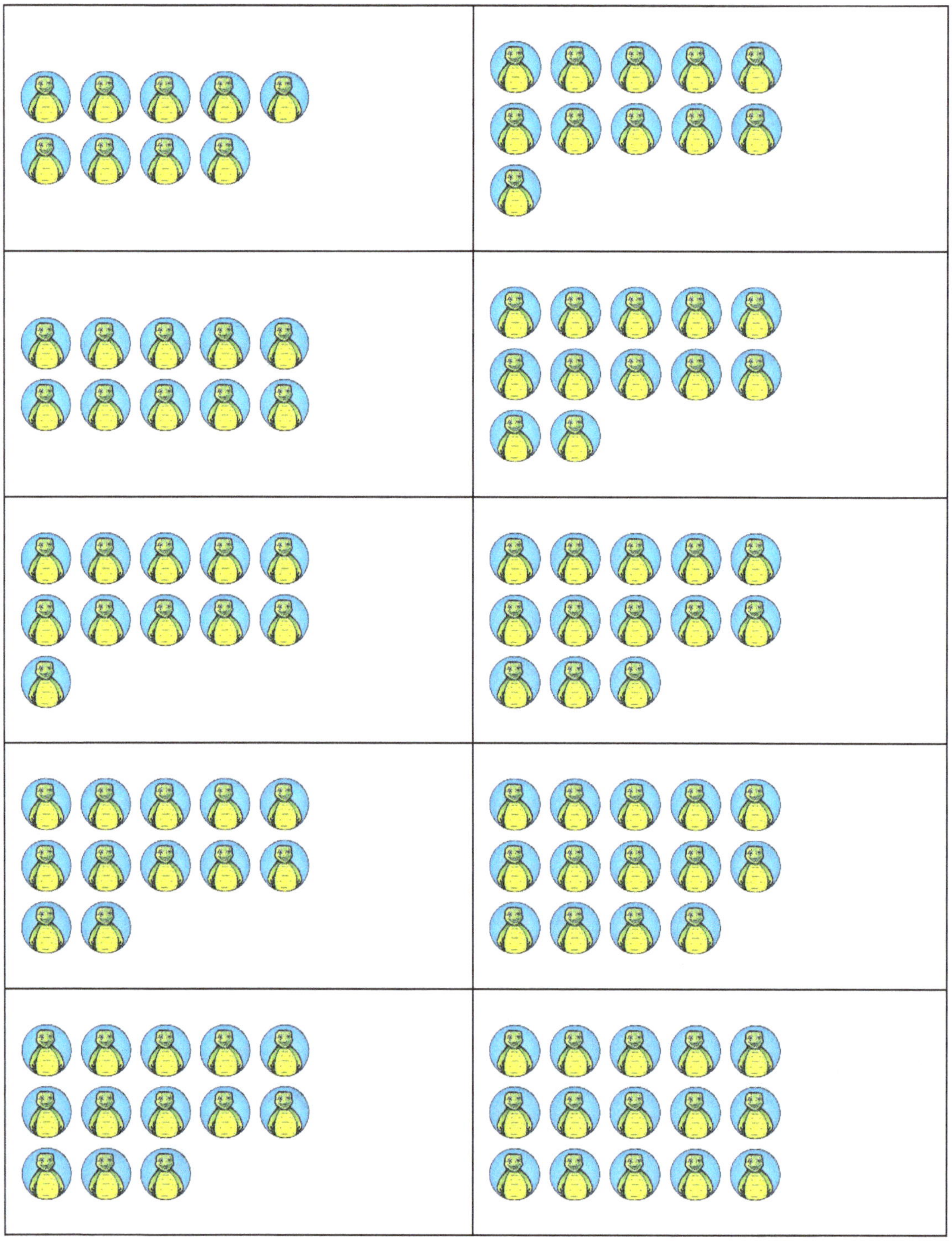

9	11
10	12
11	13
12	14
13	15

13	21
15	22
16	23
17	24
19	25

Rainbow numbers

1	2	3	4	5
6	7	8	9	10
11	12	13	14	15
16	17	18	19	20
0				

Number Track 1-20

1	2	3	4	5	glue
6	7	8	9	10	glue
11	12	13	14	15	glue
16	17	18	19	20	Cut off

Number Line 1-20

```
0    1    2    3    4    5    6    7    8    9    10
|    |    |    |    |    |    |    |    |    |    | glue

11   12   13   14   15   16   17   18   19   20
|    |    |    |    |    |    |    |    |    |
```

Ten Frames 5 wise

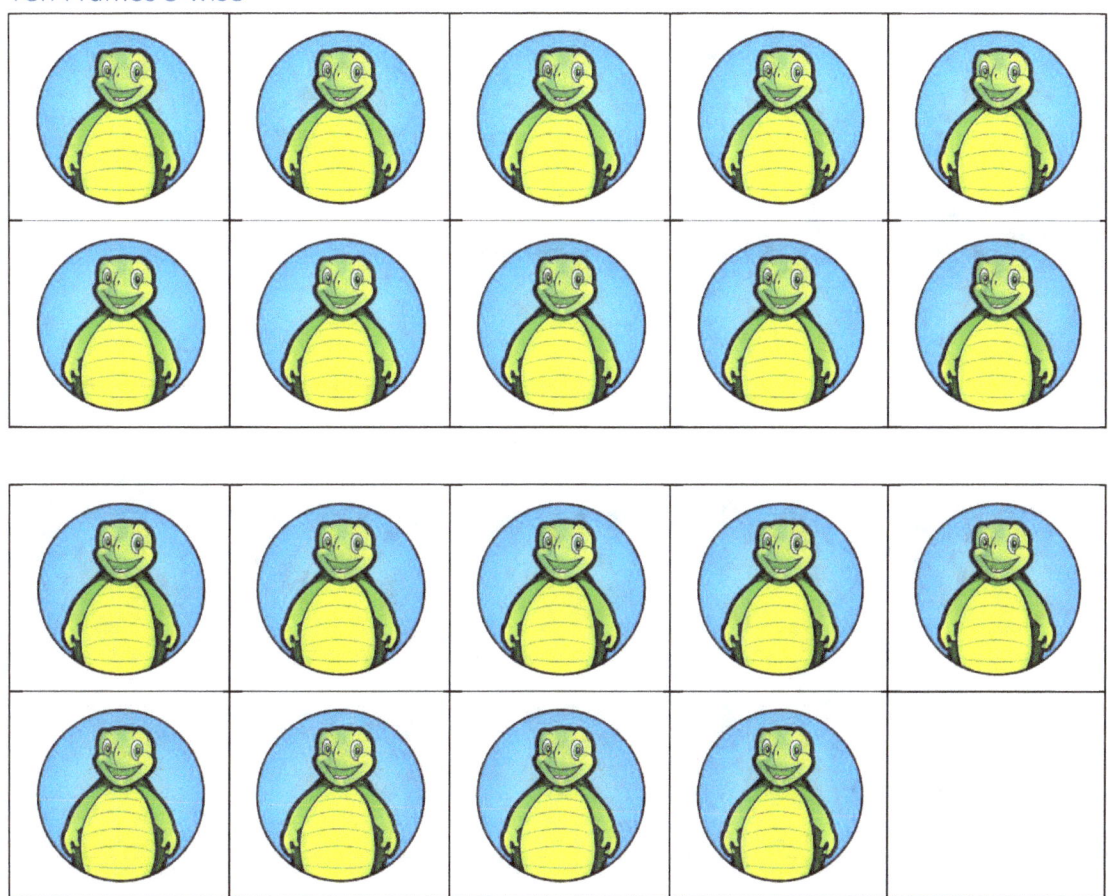

Ten frames – pair wise

Blank ten frames

Number flips – pairs to 10

Fold to show different amounts – how many on one side, how many on the other?

54

100 square

1	2	3	4	5	6	7	8	9	10
11	12	13	14	15	16	17	18	19	20
21	22	23	24	25	26	27	28	29	30
31	32	33	34	35	36	37	38	39	40
41	42	43	44	45	46	47	48	49	50
51	52	53	54	55	56	57	58	59	60
61	62	63	64	65	66	67	68	69	70
71	72	73	74	75	76	77	78	79	80
81	82	83	84	85	86	87	88	89	90
91	92	93	94	95	96	97	98	99	100

Bottom up 100 square

91	92	93	94	95	96	97	98	99	100
81	82	83	84	85	86	87	88	89	90
71	72	73	74	75	76	77	78	79	80
61	62	63	64	65	66	67	68	69	70
51	52	53	54	55	56	57	58	59	60
41	42	43	44	45	46	47	48	49	50
31	32	33	34	35	36	37	38	39	40
21	22	23	24	25	26	27	28	29	30
11	12	13	14	15	16	17	18	19	20
1	2	3	4	5	6	7	8	9	10

Go Fish Cards

1	2	3
4	5	6
7	8	9

10	11	12
13	14	15
16	17	18
19	20	0

Fingers Subitizing cards

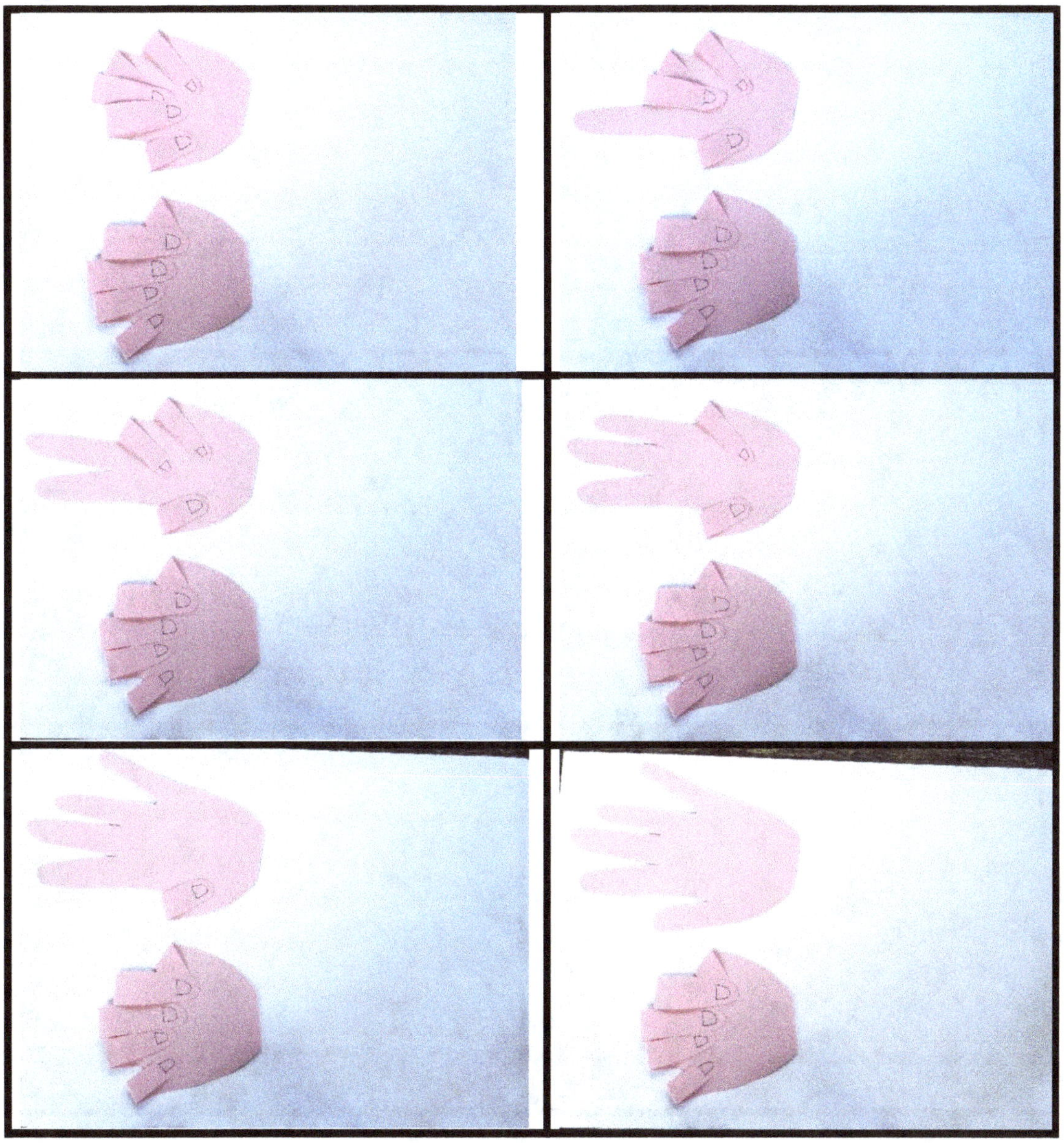

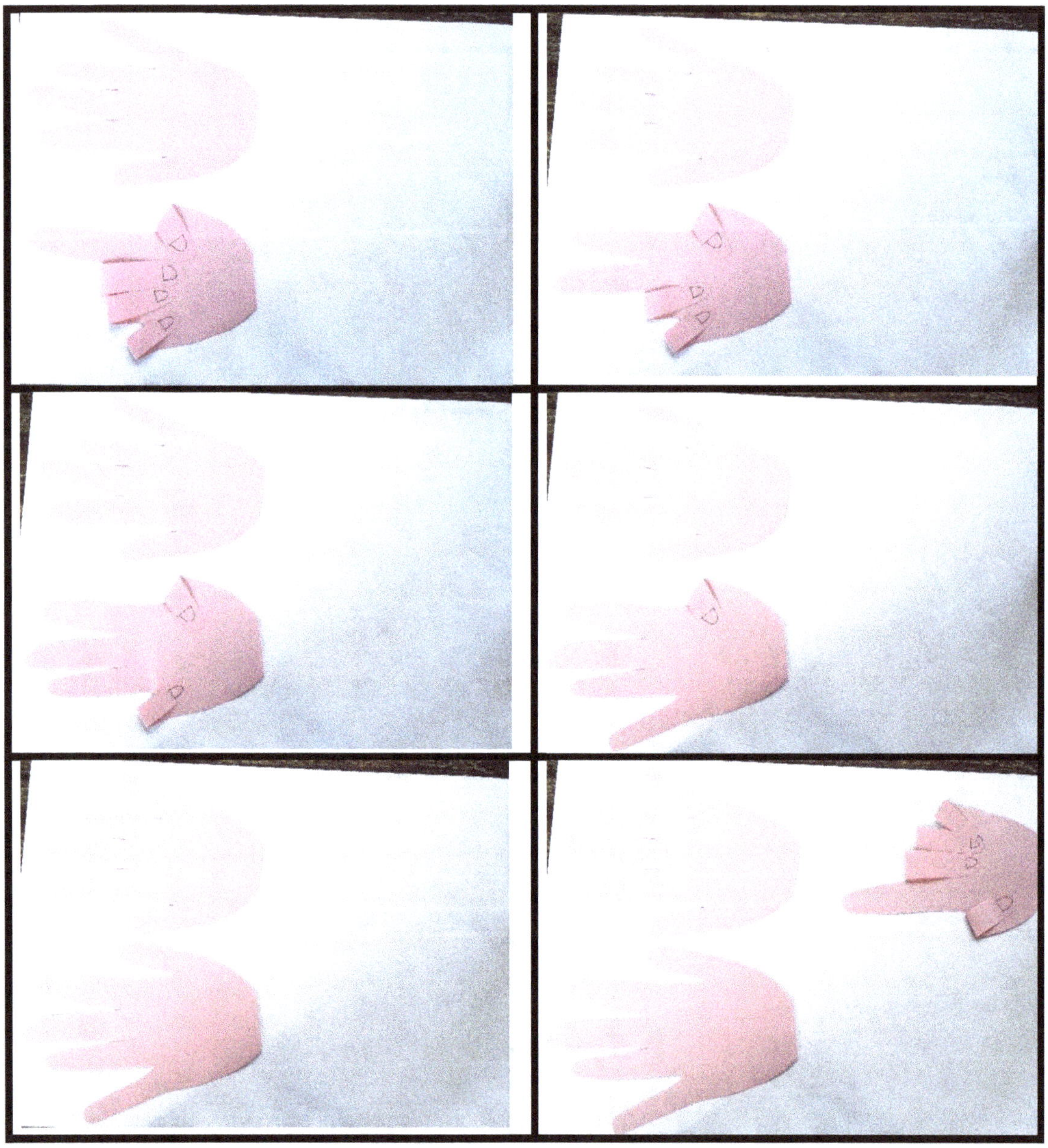

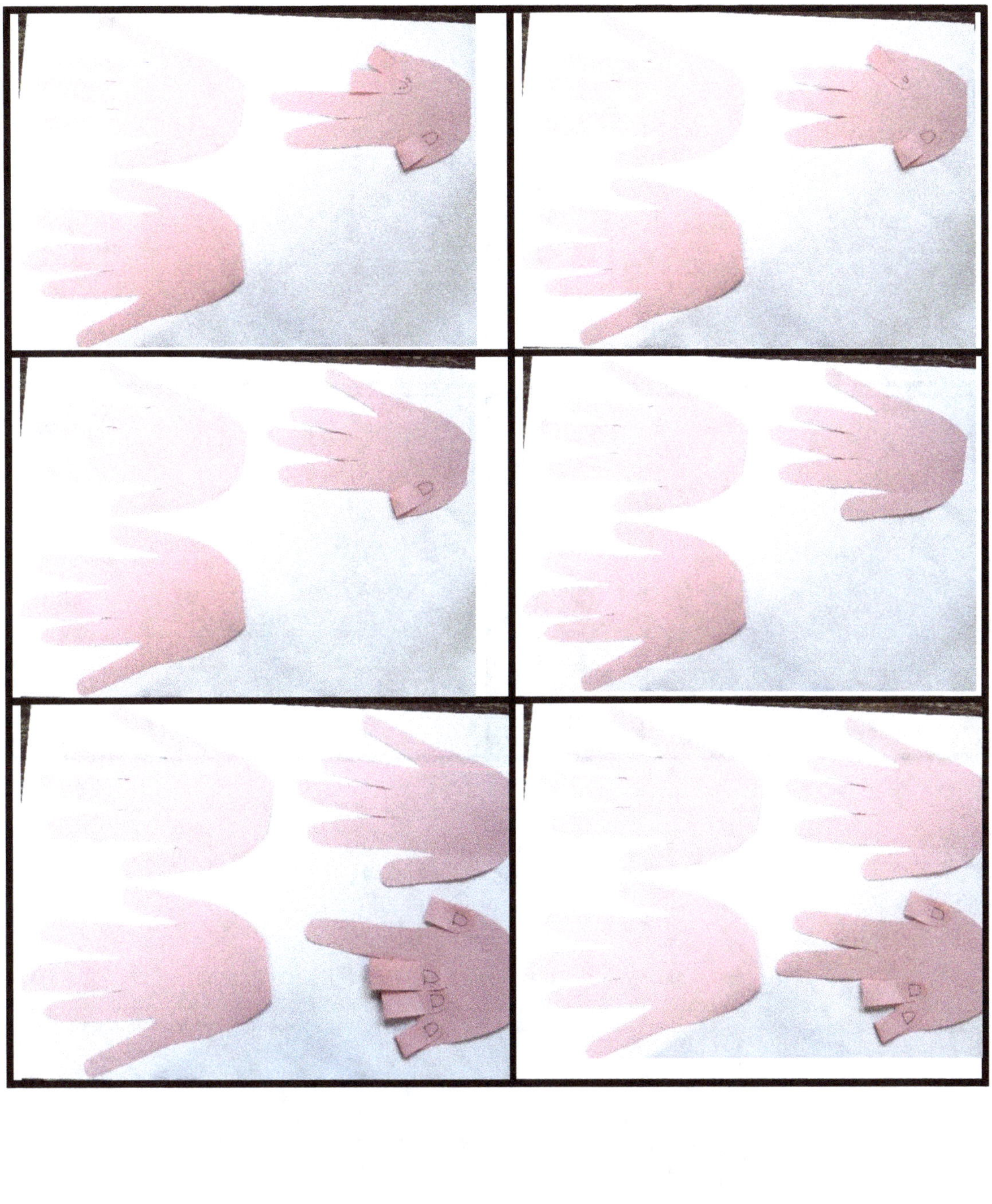

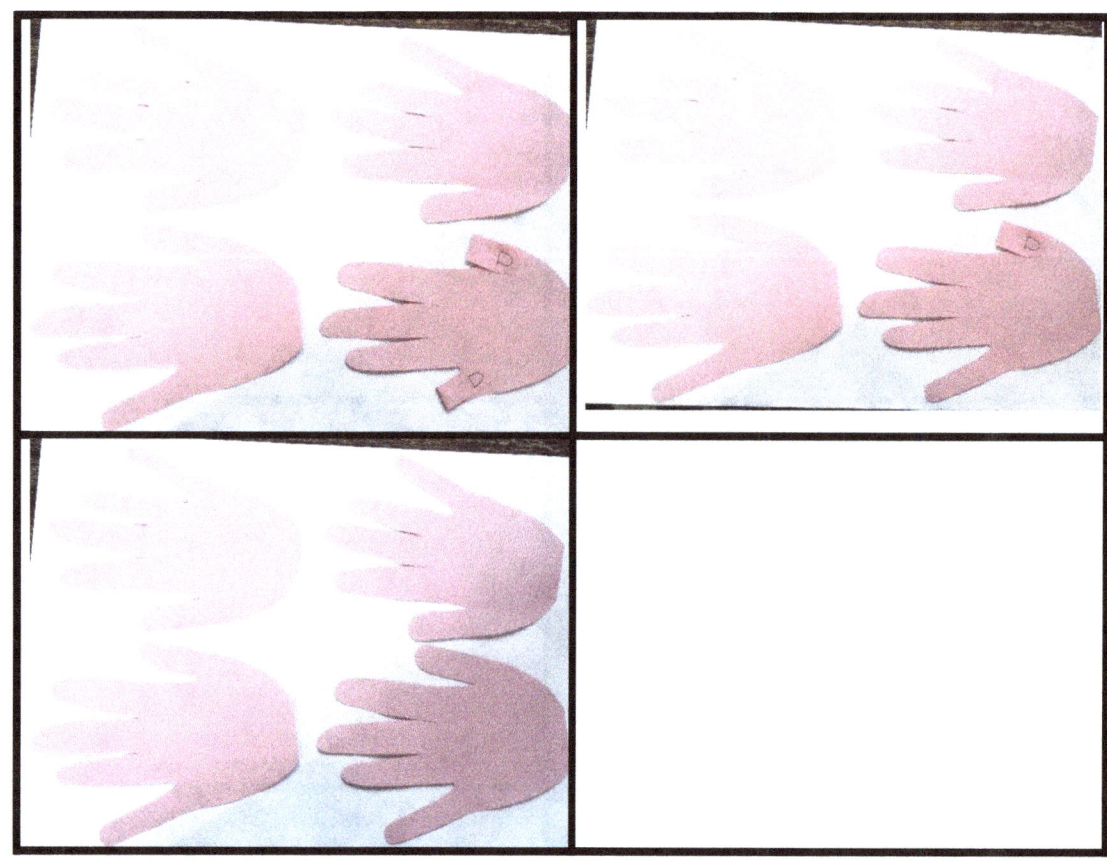

Count in 5 cards

5	10	15
20	25	30
35	40	45
50	55	60
65	70	75

80	85	90
95	100	0

Go Fish – Teen numbers

11	10+1
12	10+2
13	10+3
14	10+4
15	10+5
16	10+6

17	10+7
18	10+8
19	10+9
20	10+10

Memory game place value

(10+1)	(10+2)	(10+3)	(10+4)
(10+5)	(10+6)	(10+7)	(10+8)
(10+9)	(10+10)	11	12
13	14	15	16
17	18	19	20

Rekenrek Subitizing

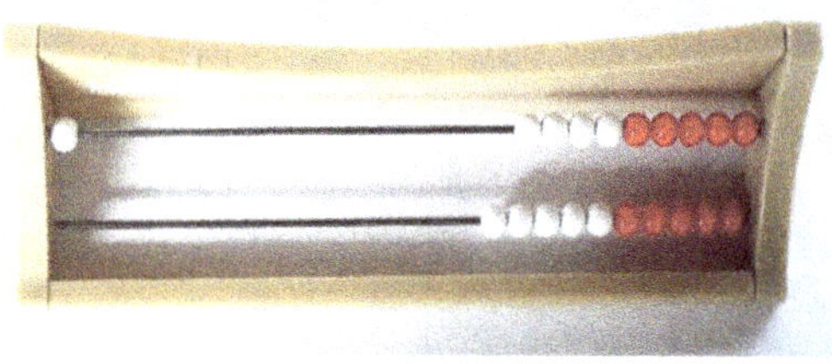

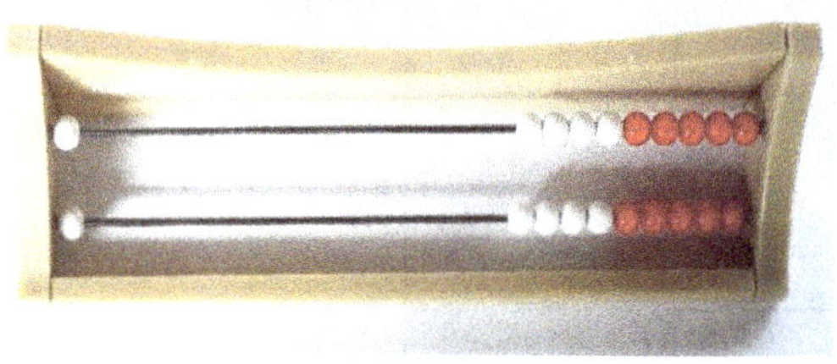

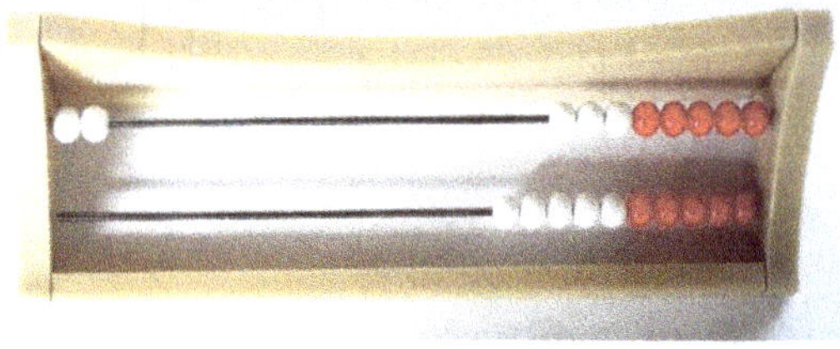

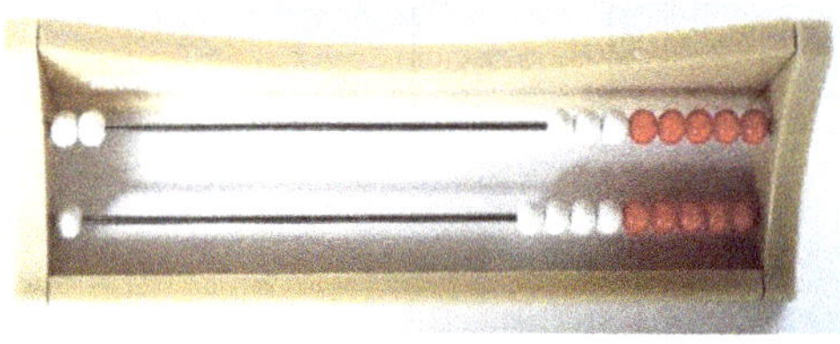

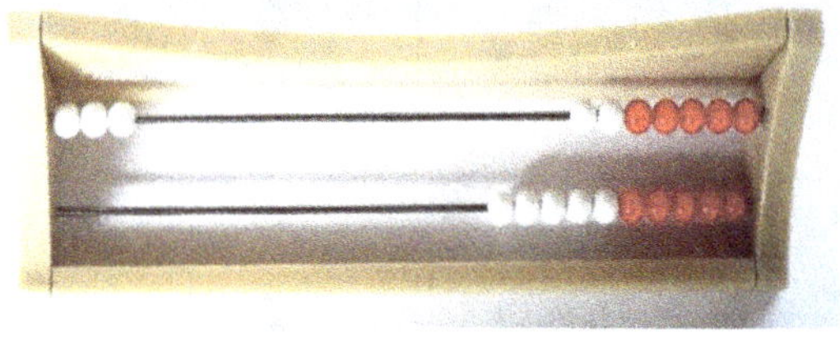

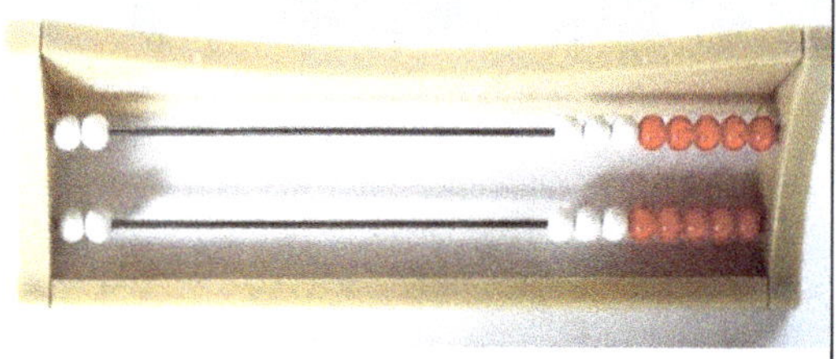

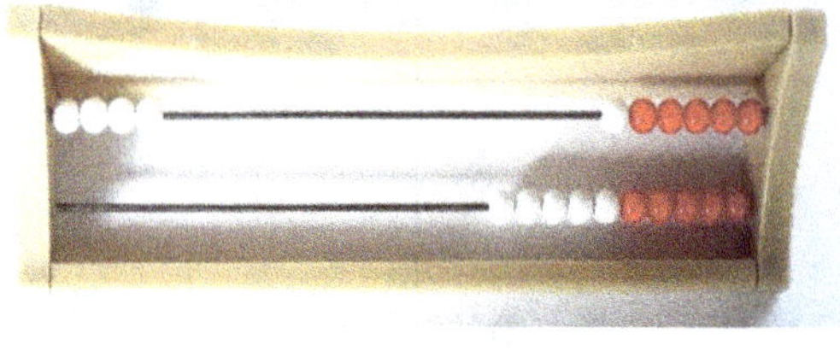

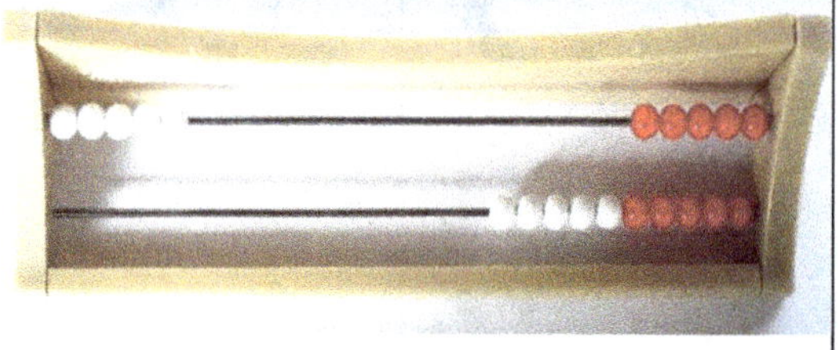

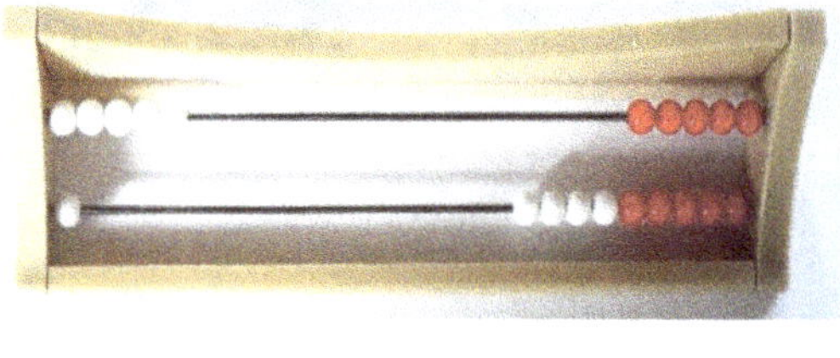

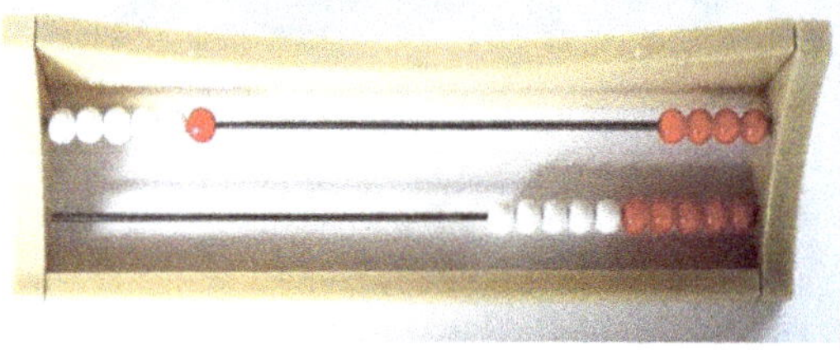

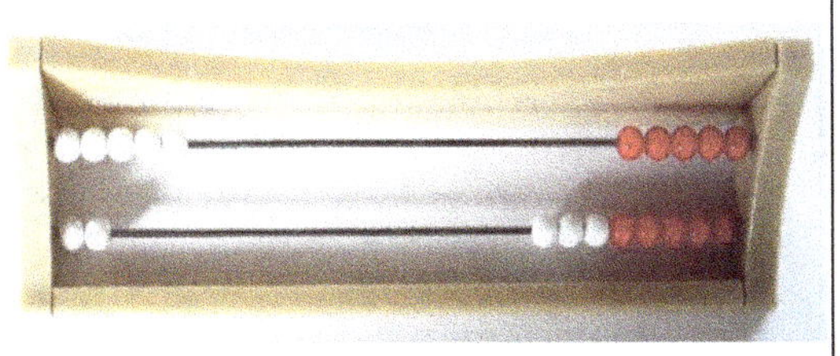

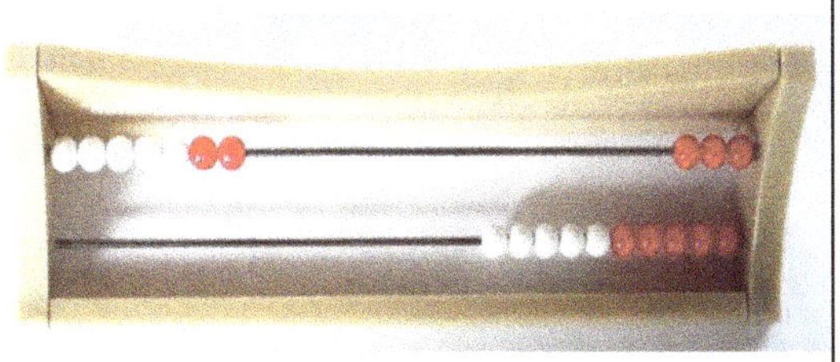

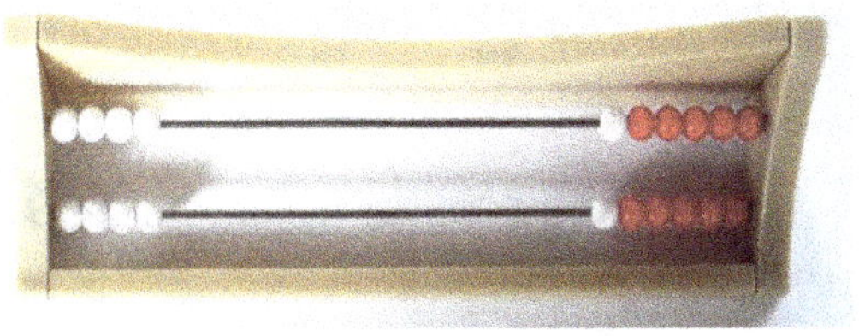

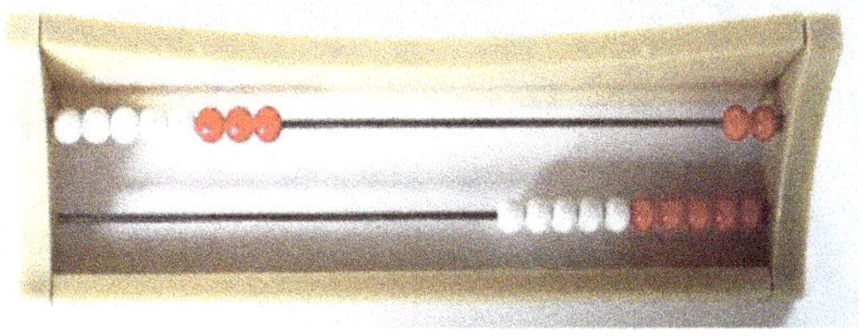

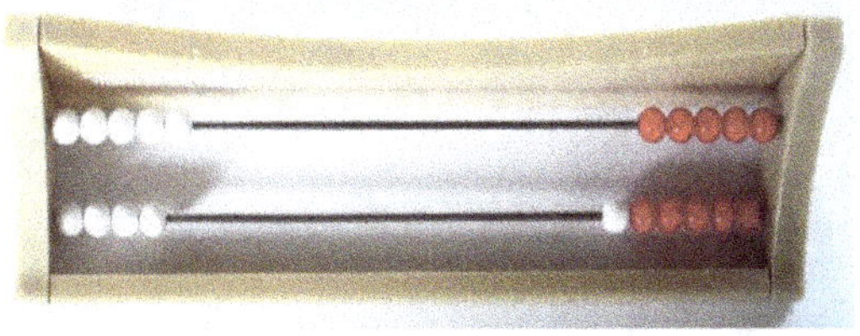

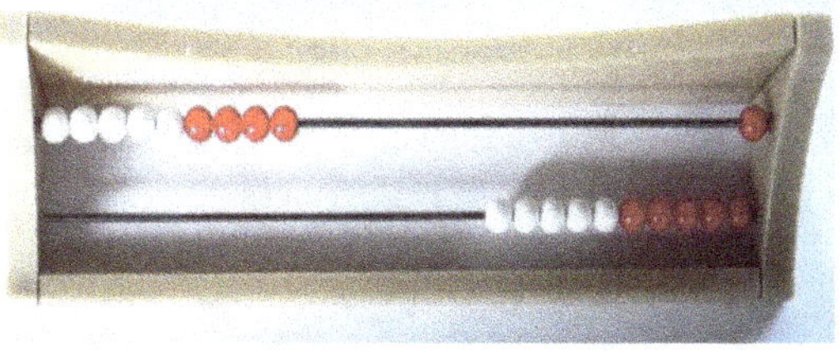

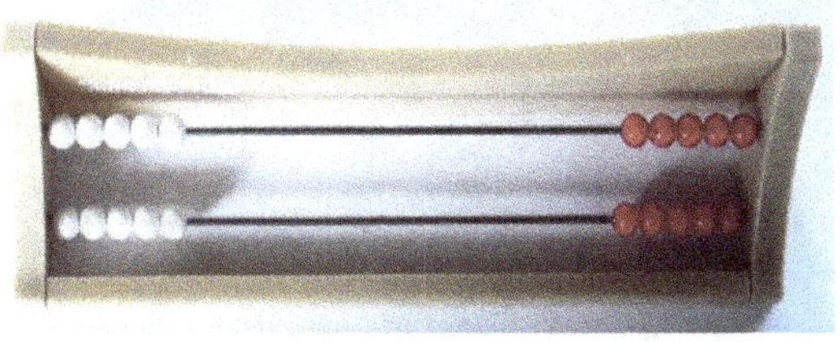

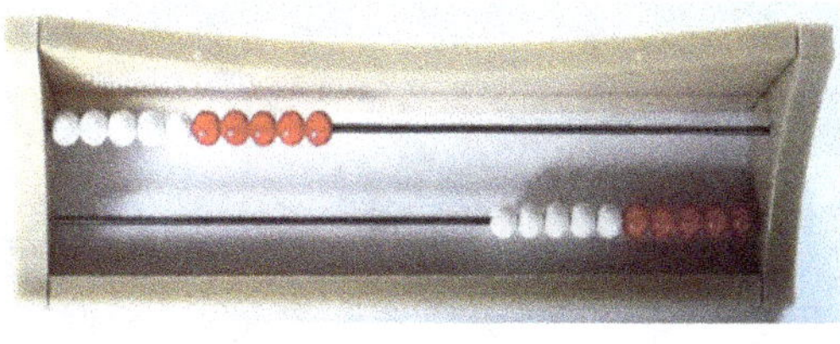

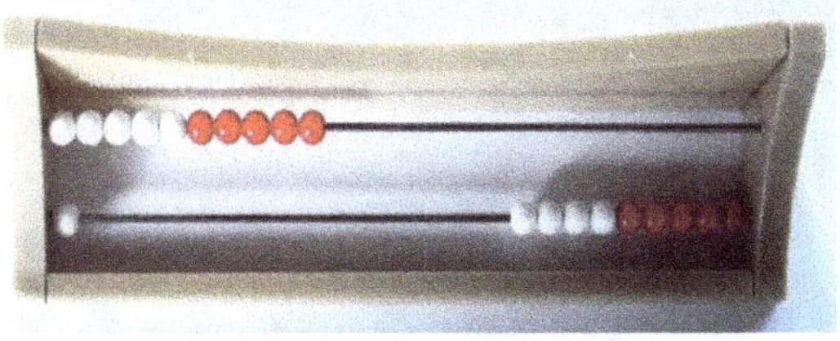

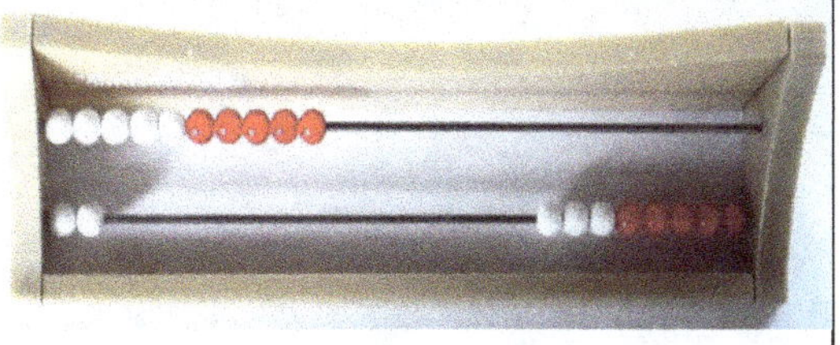

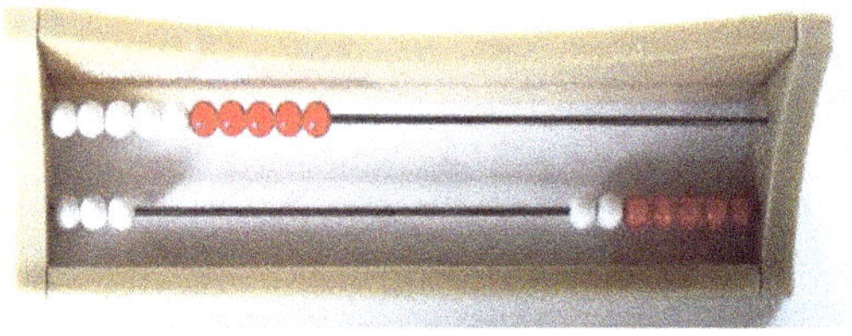

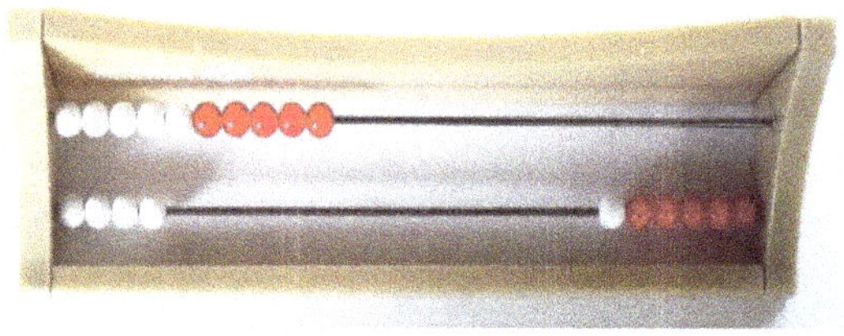

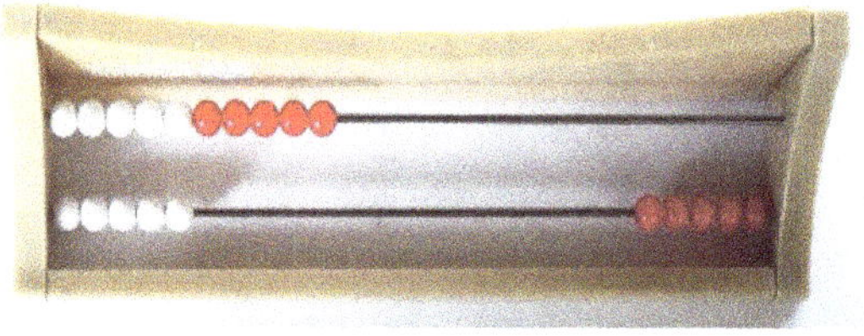

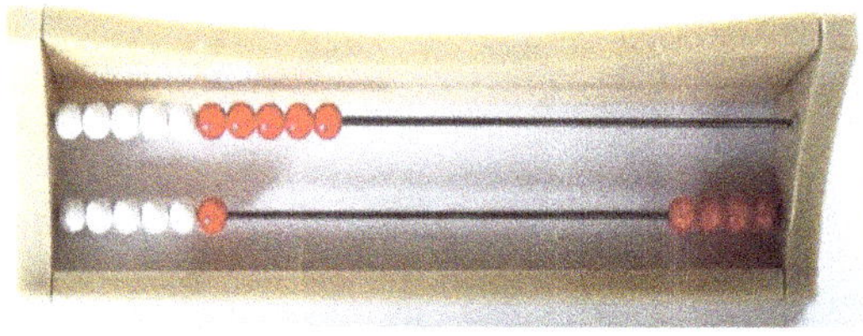

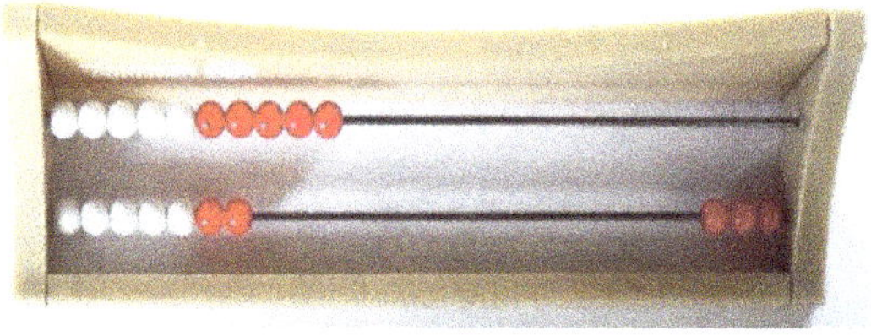

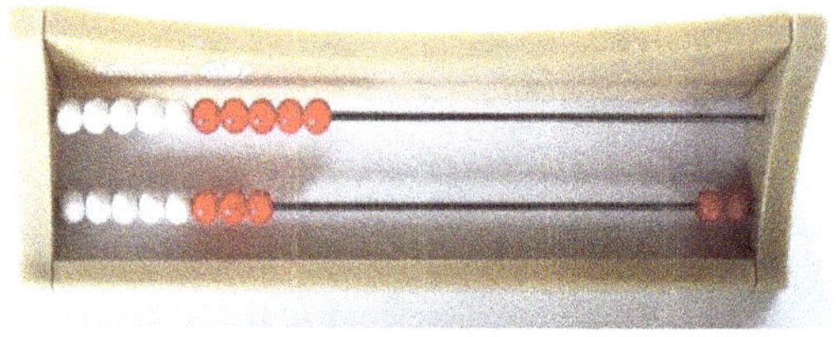

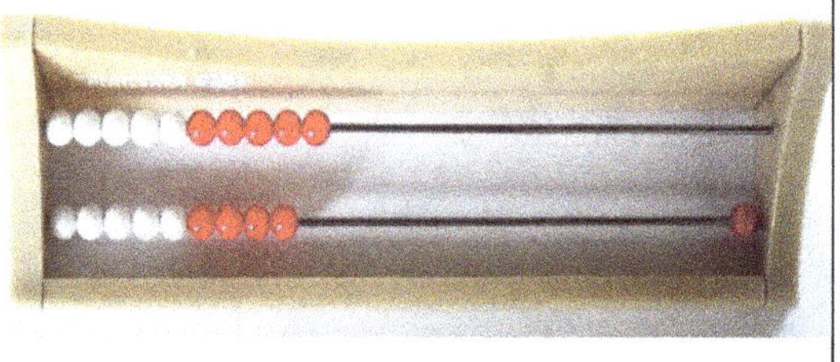

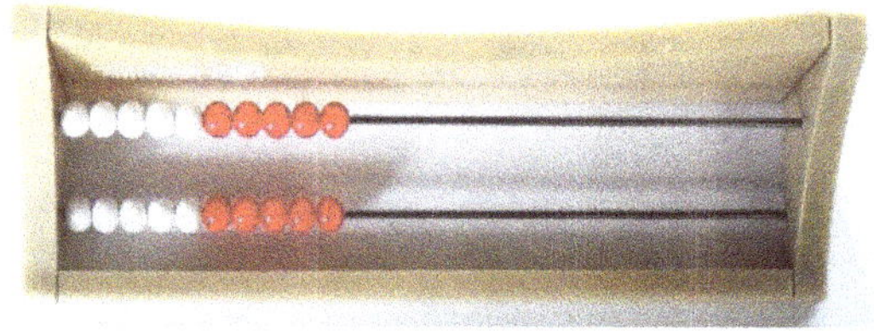

Count in 2's cards

2	4	6
8	10	12
14	16	18
20	22	24
26	28	30

32	34	36
38	40	42
44	46	48
50	0	

Doubles Bus

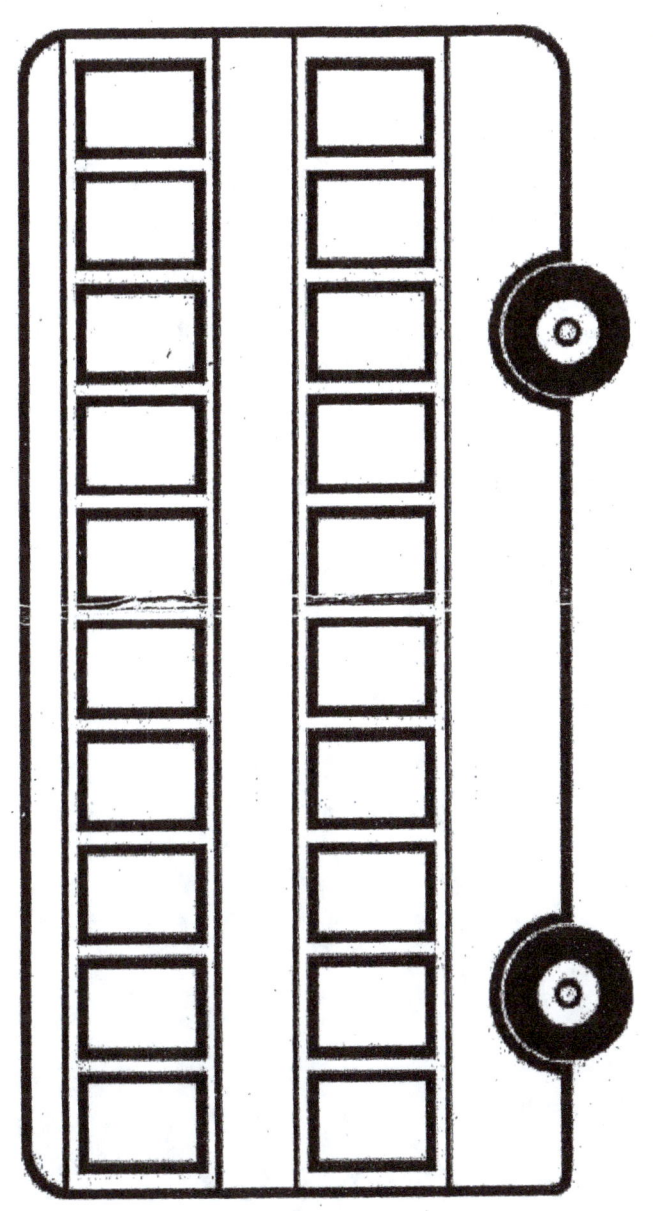

Doubles and Halves skittles game

How to play:

Doubles: write 9 even numbers from 2-10 on the skittles for the doubles game. Roll a 10-sided dice and cross off the double.

Halves: Write 9 numbers from 1-5 on the skittles for halves games. Roll 10-sided dice, roll the dice, and say what number has been halved to get the answer e.g., if you roll 6, that is half of 12.

Near Doubles: write 9 numbers from 1,3,5,7,9,11. Roll a 6-sided dice and cross of a near double e.g., roll 3 can cross off 5 or 7 (double 3 is 6 so near double is 5 or 7)

Winner is the first to cross off all skittles.

If you roll a 6 or *, roll again.

Miss your go if you cannot cross off a skittle.

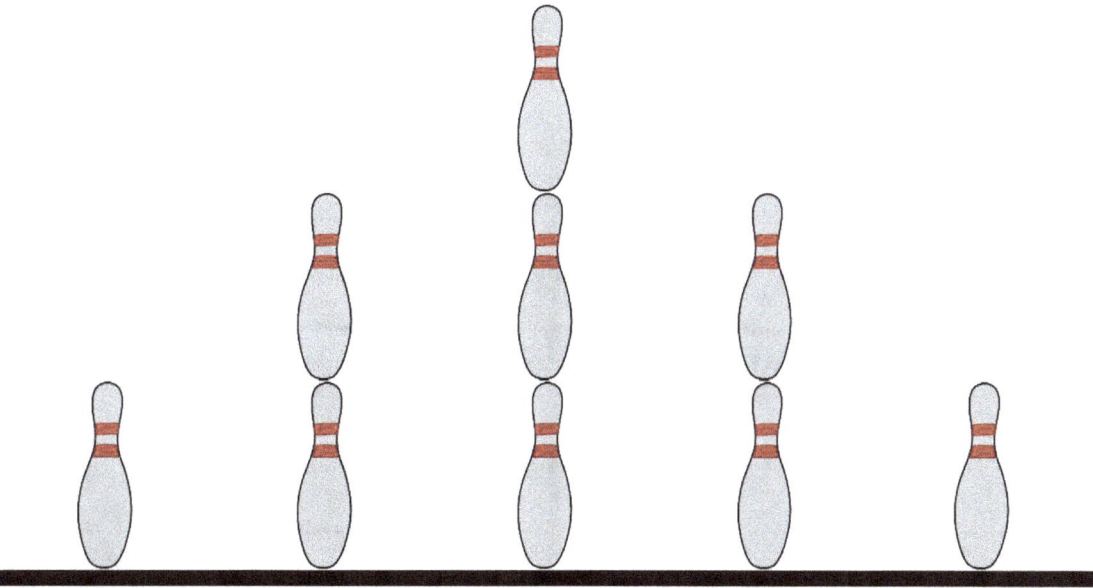

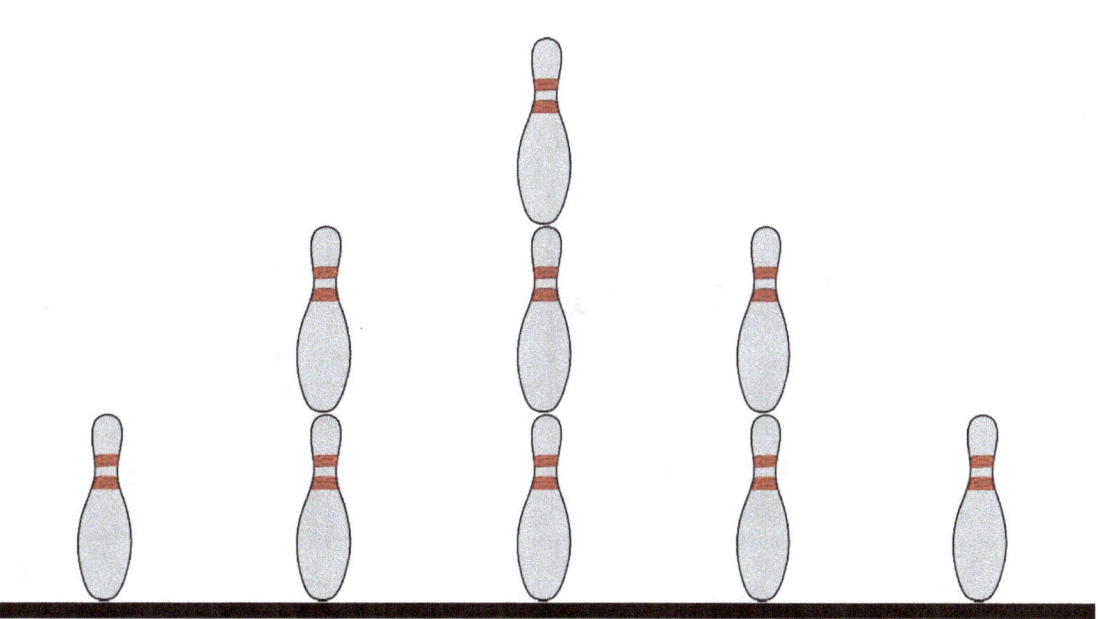

Follow on doubles and halves game

Start with the start card and answer the question and place down the next card. Doubles cards are blue and must follow blue. Halves are black and follow black. The game end when the stop card is played, and all cards have been played in the right order.

To cut out the cards, use the bold black line to separate the 2 different cards in each row. Then cut across.

Start	Double 1	2	Double 5
10	Half of 14	7	Double 3
6	Half of 16	8	Half of 4

2	Double 8	16	Half of 2
1	Half of 10	5	Double 6
12	Double 7	14	Half of 20
10	Half of 18	9	Double 9

18	Double 2	4	Half of 8
4	Double 4	8	Double 10
20	Half of 6	3	Half of 12
6	Stop		

Number flips – doubles and near doubles

Cut across the page and then fold down the middle so you can open and compare if needed.

2 penguins	3 penguins
2 penguins	5 penguins
3 penguins	7 penguins
3 penguins	6 penguins
4 penguins	10 penguins

4	8
5	11
5	10
6	13
6	11

7 penguins	13 penguins
7 penguins	15 penguins
8 penguins	22 penguins
8 penguins	20 penguins
9 penguins	19 penguins

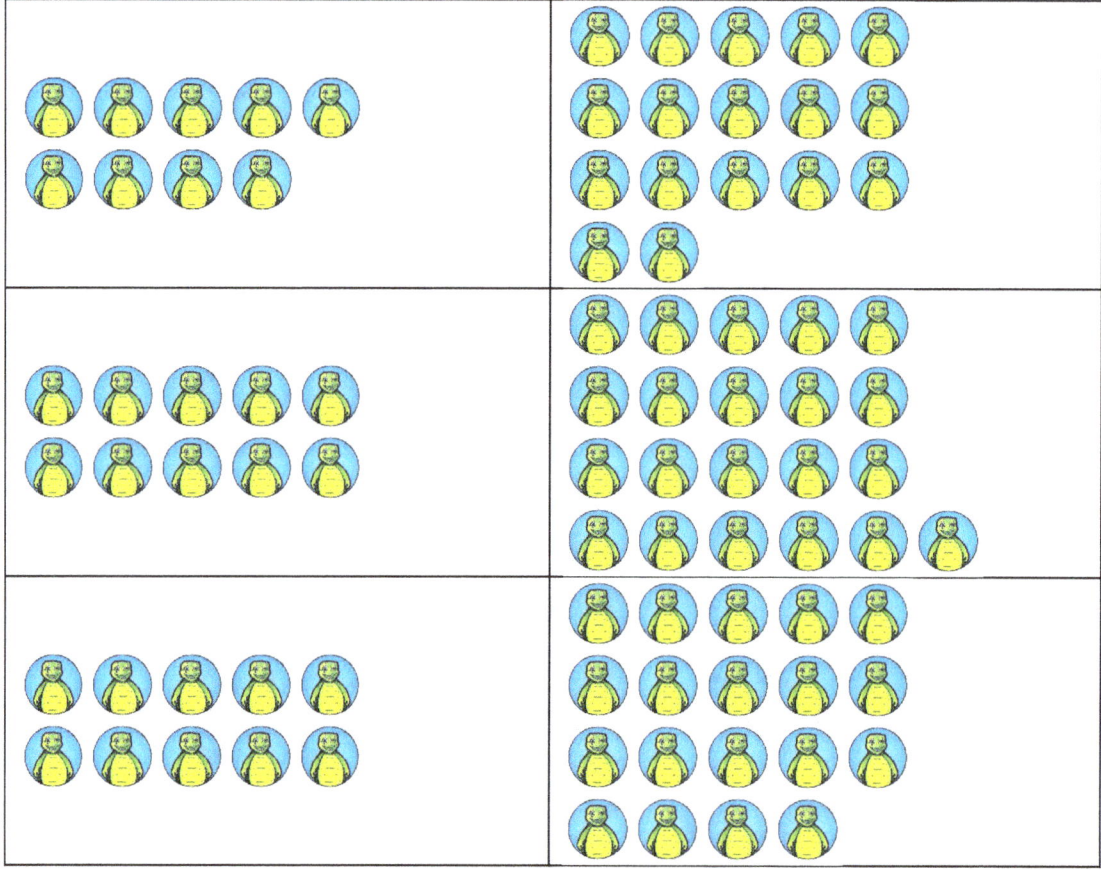

Near Doubles go fish or memory game

0+1	1
1+1	2
1+2	3
2+2	4

2+3	5
3+3	6
3+4	7
4+4	8

4+5	9
5+5	10
5+6	11
6+6	12

6+7	13
7+7	14
7+8	15
8+8	16

8+9	17
9+9	18
10+9	19
10+10	20

Counting in 10's cards

10	20	30
40	50	60
70	80	90
100	0	

Arrow cards
Cut out the cards and make the arrows pointed as shown.

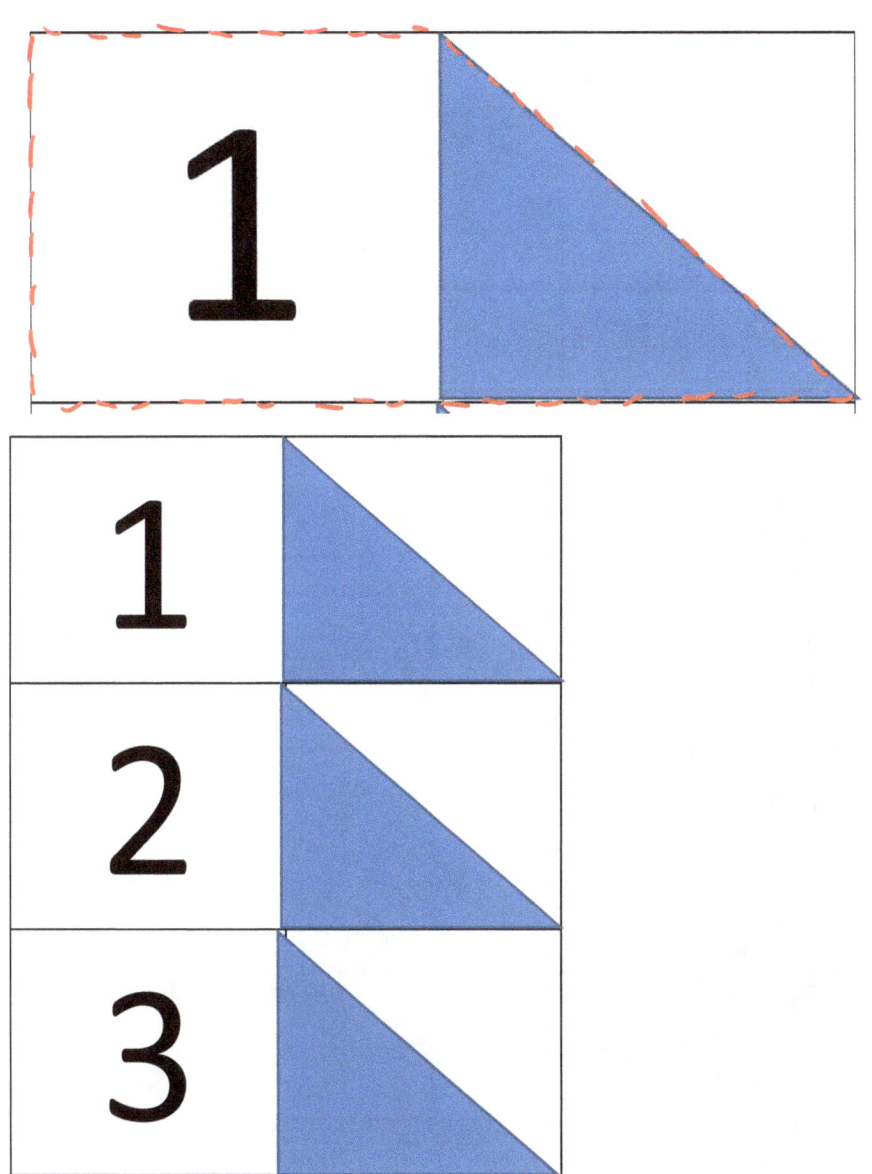

4	
5	
6	
7	
8	
9	

1	0	
2	0	
3	0	
4	0	
5	0	
6	0	

7	0	◣
8	0	◣
9	0	◣

Number track 10-20

10	20	30	40	50	glue
60	70	80	90	100	glue
110	120				

Tally mark subitizing

|	||
|||	||||
⋕	⋕ |
⋕ ||	⋕ |||

𝍳𝍳𝍳𝍳𝍳 𝍳𝍳𝍳 (9)	𝍳𝍳𝍳𝍳𝍳 𝍳𝍳𝍳𝍳 (9)
𝍳𝍳𝍳𝍳𝍳 𝍳𝍳𝍳𝍳𝍳 𝍳 (11)	𝍳𝍳𝍳𝍳𝍳 𝍳𝍳𝍳𝍳𝍳 𝍳𝍳 (12)
𝍳𝍳𝍳𝍳𝍳 𝍳𝍳𝍳𝍳𝍳 𝍳𝍳𝍳 (13)	𝍳𝍳𝍳𝍳𝍳 𝍳𝍳𝍳𝍳𝍳 𝍳𝍳𝍳𝍳 (14)
𝍳𝍳𝍳𝍳𝍳 𝍳𝍳𝍳𝍳𝍳 𝍳𝍳𝍳𝍳𝍳 (15)	𝍳𝍳𝍳𝍳𝍳 𝍳𝍳𝍳𝍳𝍳 𝍳𝍳𝍳𝍳𝍳 𝍳 (16)

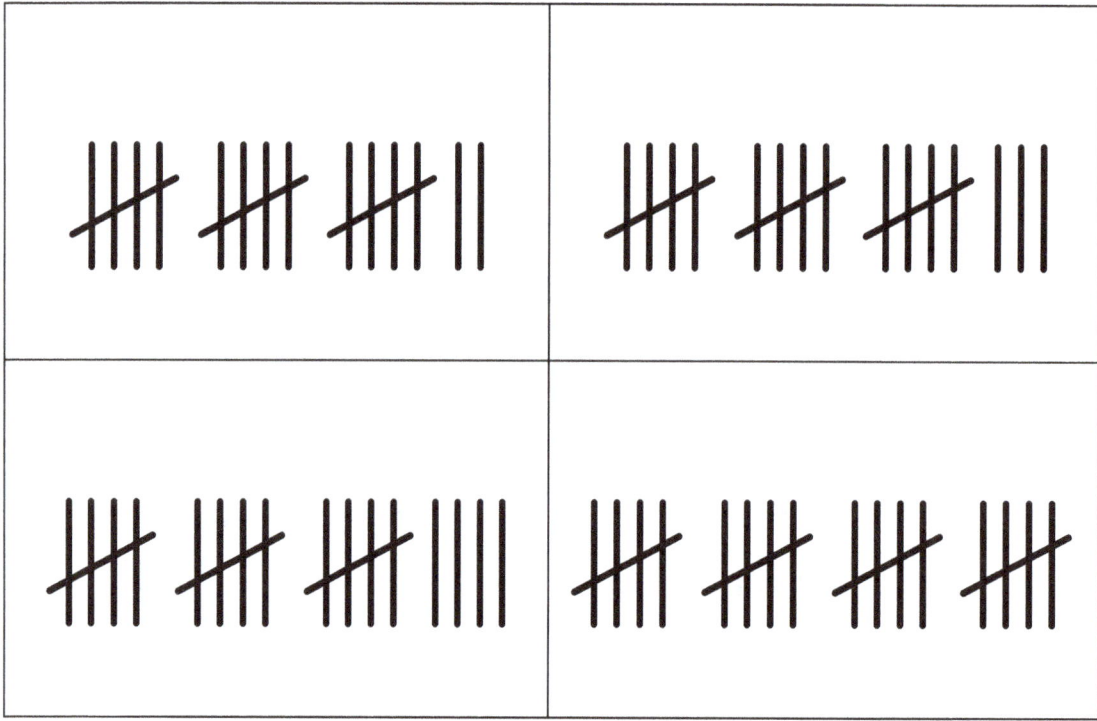

Pocket Game

Print 1 copy for the game board (works well if you enlarge this for the game board) and 2 more to cut up for the playing cards.

1	2	3	4	5
6	7	8	9	10
11	12	13	14	15
16	17	18	19	20

Level 2 Numbers to 20 Workbook

The photocopiable pages, number problems booklet and word problems are available at:
http://mathtastic.com.au/mathtastic-resources-level-2/
Password: Mathtastic2

Contents

- Level 2 Numbers to 20 99
- Workbook 99
- Instructions 101
- Setting out the student book 102
- Coding the answers 104
- Module 1 105
  - Draw and solve 105
  - Draw and Solve 105
- Module 2 106
  - Draw and solve 106
  - Retrieval Practice 106
- Module 3 107
  - Draw and solve 107
  - Retrieval Practice 107
- Module 4 108
  - Draw and solve 108
  - Retrieval Practice 108
- Module 5 109
  - Draw and solve 109
  - Retrieval Practice 109
- Module 6 110
  - Draw and solve 110
  - Retrieval Practice 110
- Module 7 111
  - Draw and solve 111
  - Retrieval Practice 111
- Module 8 112
  - Draw and solve 112
  - Retrieval Practice 112

Instructions

Answer the questions on each page. Each row is for a different day. The first column contains practice related to the work in the module and the second column is for practice from previous modules. Problems should be presented alternatively horizontally and vertically from Module 2 onwards.

Students (or adults) should copy the problem into a maths exercise book – this can be plain, lined or squared. In the early stages a plain book may be easier.

Divide the page into 2 columns – the first for writing the problem and the second to represent in the 5 different ways. See below for an example.

For each of the questions in the first column solve the problem and note how it was solved in the third column – see below. Then show understanding using one or more of the ways shown on the conceptual understanding pentagon below.

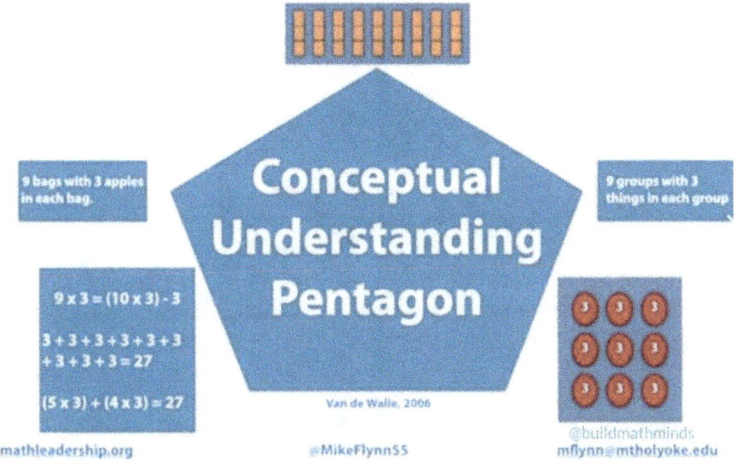

This way students can show their conceptual understanding or misunderstanding:

- Model – show on a model e.g., ten frame, Rekenrek, numicon
- Words – write out in works – six plus three equals nine.
- Pictures – draw out the problem as a picture
- Equations – write the equation (already done for you but you could rewrite vertically or horizontally)
- Contexts – write a number story – e.g. I had 3 apples, and I ate 1. Now there are only 2 apples.

Setting out the student book

$$6 + 2 = 8$$

Six plus two equals eight.

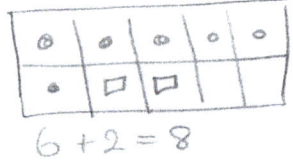

Jane had 6 puppies and her Poppy brought her 2 more. Now she has 8 puppies!

6 + 2 = 8

$4 - 1 = 3$

Four subtract one equals three.

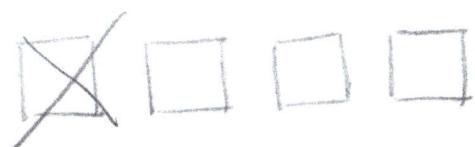

Four birds were in the garden. One flew away. Now there are three birds

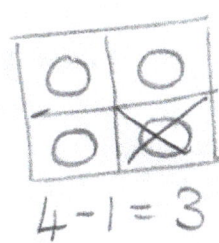

$4 - 1 = 3$

Coding the answers

Coding the answers will help to diagnose which strategies the students are using and which they are not. We really need students to move beyond counting in ones and seeing numbers in bigger groups as well as learning some of these facts so they can recall them automatically.

Coding the student answer strategy: ask the student how they worked out the answer – they may use more than one way	
Automatic – just knew the answer, immediate recall	A
Counted on or back a small number (+/- 0,1,2,3)	S
Rainbow facts – knew the pair made 10	RF
Counted on from largest number	CO
Counted back for subtraction	CB
Doubled or halved	D or H
Near double	ND
Place value – e.g., 10+4= 14	PV
Compensated – made an adjustment to the number before calculating e.g., 9+6 – rearranged to 10+5 as easier	C
Other – you may wish to note this down	O

Module 1

Draw and solve	Code	Draw and Solve	Code
19+0= 14-1 = 16+2 = 17-2= 15+3=		12+2= 17-0= 13+3= 14-2= 11-3=	
18-1= 14+2= 17-0= 20-2= 16+3=		11+3= 12+3= 19-2= 18-0= 17-3=	
15-0= 14+3= 15+2= 16-2= 17-1=		13-2= 17+3= 18+0= 20-3= 19+1=	

Module 2

Draw and solve	Code	Retrieval Practice	Code
5+12= 12-8= 19-4= 13+5= 15-?= 12 16+?=19		17-0= 1+18= 15-3= 3+17= 18-1= 14+?=15 19-?=17	
7+12= 15+1= 18-5= 16-4= 13-?=9 14+?=17		16-1= 17+2= 17-1= 2+15= 15+?=17 14-? =11	
16+3= 8+11= 20-7= 15-3= 12-?=8 17+?=20		15-2= 15+3= 3+16= 18-2= 17-?=19 12-?=7	
11+6= 4+12= 18-4= 20-4= 15-?=11 17+?=19		14-3= 0+15= 19-3= 18-3= 8+2= 17-?=14 16+?=19	

Module 3

Draw and solve	Code	Retrieval Practice	Code
Solve – is it a rainbow fact to 20? Find the odd one out. 10+10= 15+5= 8+12= 7+11= 16+?= 20 13+?= 20		14+1= 16-2= 18-5= 5+12= 6+6= 17-?=16 13+?=17	
7+13= 16+2= 9+11= 5+15= 13+?=20 8+?=20		18-3= 5+13= 7+8= 5+14= 17-?=15 15+?=19	
12+8= 3+17= 4+16= 19+2= 14-?=12 17+?=20		14+2= 17-0= 19-16= 11+8= 15+?=19 18-?=6	
11+6= 15+5= 1+19= 6+14= 18-?=14 20-?=15		18-1= 19+0= 18-6= 12+7= 18-?=12 15-?=13	

Module 4

Draw and solve	Code	Retrieval Practice	Code
15-5= 10+5= 10+9= 17-10= 18-?=10 16_?=6		16+3= 17+?=20 20-14= 6+12= 16-5= 8-?=4	
18-10= 5+10= 16-10= 16-6= 13-?=10 14-?=4		19-3= 18+2= 20-?=15 14-10= 20-7= 14+?=19	
17-10= 4+10= 10+8= 19-9= 12-?=2 15-?=10		17+2= 16+?=20 20-6= ?+8=19 18-17= ?+3=16	
16-6= 11-10= 14-4= 20-10= ?-9=10 ?-10=7		20-2= 19+1= 20-?=12 0+20= 17-4= ?+13=17	

Module 5

Draw and solve	Code	Retrieval Practice	code
6+6= 7+7= 8+8= 9+9= 18-?=9 ?-7=7		17+1= 17-6= 18+2= 17-5= ?-10=8 6+10=	
0+0= 14-7= 6+?=12 10-5= 16-8= 8+?=16		18-2= 12+6= 17+3= 15+3= 5+10= 17-?=10	
18-9= 9+9= 12-6= 7+7= 16-?=8 10-?=5		19+0= 11+7= 16+4= 19-5= 19-?=10 13-?=3	
6+6= 0+0= 10+10= 14-?=7 18-?=9 9+9=		16+3= 18-5= 15+5= 20-5= 12-?=2 20-?=20	

Module 6

Draw and solve	Code	Retrieval Practice	code
5+4= 5+6= 8+7= 9+8= 17-8= ?+6=13		16+2= 17+3= 18-5= 12+2= 12-?=7 ?+5=10	
6+5= 10+9= 8+7= 7+8= 13-?=7 ?+8=15		19-7= 14+6= 19-6= 18-4= 13+?=16 ?-12=2	
5+4= 5+6= 9+8= 9+10= 17-?=9 ?-5=6		17-3= 15+5= 15-15= 16-3= 14+?=18 ?=9=9	
4+5= 6+5= 8+9= 7+8= 15-?=8 ?-9=8		13+6= 11+9= 17-5= 14-2= 15+?=20 ?+6=11	

Module 7

Draw and solve	Code	Retrieval Practice	Code
At this level, this module is the same as module 4. 10+9= 8+10= 10+6= 7+10= ?-3=10 14-?=10		16+0= 20-17= 8+8= 8+9= 6+6= 6+7= ?-14=14	
2+10= 10+2= 10+5= 4+10= 6+10= 17-?=10		11+6= 15+5= 15+4= 18-5= 16-3= 14+3=	
10+3= 10+7= 1+10= 3+10= 9+10 18-?=10		7+7= 7+8= 17-2= 20-4= 18-4= ?+2=15	
5+10= 10+4= ?+8=18 10+1= 10+10= 4+?=14		9+9= 9+8= 19-7= 15+?=17 15+5= ?-4=14	

Module 8

Draw and solve	Code	Retrieval Practice	Code
9+6= 8+5= 7+6= 9+4= 7+5= 8+4=		12+4= 11+3= 17+3= 19-7= 13-1= 19-8=	
9+7= 5+6= 6+8= 7+9= 3+8= 2+9=		9+9= 9+8= 14-3= 18-6= 18-1= ?-16=0	
9+4= 13-?=9 12-?=8 15-6= 13+6= 14+8= 16-?=9		16+1= 9-2= 8-2= 12-12= ?-6=4 5+3= 15-?=5	
8+4= 6+8= 14-?=8 ?-6=7 9+2= 8+5=		3+3= ?+6=12 ?+7=15 16-8= 11-?-10 19-10=	

Level 2

Numbers to 20

Problem Solving book

The photocopiable pages, number problems booklet and word problems are available at:
http://mathtastic.com.au/mathtastic-resources-level-2/

Password: Mathtastic2

Contents

Level 2 113

Numbers to 20 113

Problem Solving book 113

    How to use these cards 115

    Recording Page 116

    Module 1 – Add or subtract 0,1,2,3 117

    Module 2 – Add from the largest number, subtract by counting back 124

    Module 3 – Rainbow facts, pairs of numbers to 20 131

    Module 4- Add and subtract 10 138

    Module 5- Doubles to 20 145

    Module 6 – Near doubles.. 152

    Module 7 – Add using place value 159

    Module 8 – Add and subtract by compensation strategies 160

How to use these cards

How to use these problem cards

Each card is written to meet each of the different ways of presenting a problem. These are written at the bottom of each page for teacher reference.

The problems are deliberately written with each piece of the problem written on a separate line. This way you can cover up the rest of the problem and reveal one line at a time.

Bet lines- this is a technique to help students to think about the problem. You show one line at a time and as students to "bet" what the problem is going to ask them to do. They review their ideas as more information is revealed.

For each problem there are 2 extra sets of numbers which can be used instead for extra practice of that problem type -at the bottom of the page in brackets next to the problem type)

Recording Page

Use this page to record how the student managed with each different problem type. Which can they do easily, and which need more practice?

Join – result unknown 6+2=?	Join – change unknown 6+?=8	Join – start unknown ?+2=8
Separate – result unknown 9-5=?	Separate – change unknown 9-?=4	Separate – start unknown ?-5=4
Part-part-whole – whole unknown 5+4=?	Part- part-whole – part unknown ?+4=9	
Compare – difference unknown 7-2=?	Compare – compared set unknown 7-?=5	Compare – referent unknown ?-5=2

Module 1 – Add or subtract 0,1,2,3

Georgie had 15 crayons.
Sam gave her 2 more for her birthday.
How many crayons does she have now?

Join – result unknown (18/1, 14/3, 16, 1)
Mathtastic Level 2 – Numbers to 20
Add and subtract 1 or 2 to 20
tracyashbridge.com

Georgie had 16 crayons.
She got some more for her birthday.
Now she has 18 crayons.
How many crayons did she get for her birthday?

Join – change unknown (15/17, 12/15, 14/15)
Mathtastic Level 2 – Numbers to 20
Add and subtract 1 or 2 to 20
tracyashbridge.com

Georgie had some crayons.
Her friend gave her 1 more.
Then she had 12 crayons.
How many crayons did she have at the beginning?

Join - start unknown (2/14, 1/17, 3/13)
Mathtastic Level 2 - Numbers to 20
Add and subtract 1 or 2 to 20
tracyashbridge.com

Sam had 15 pencils.
She gave 1 to Sarah.
How many pencils does Sam have left?

Separate - result unknown (14/2, 18/1, 17/2)
Mathtastic Level 2 - Numbers to 20
Add and subtract 1 or 2 to 20
tracyashbridge.com

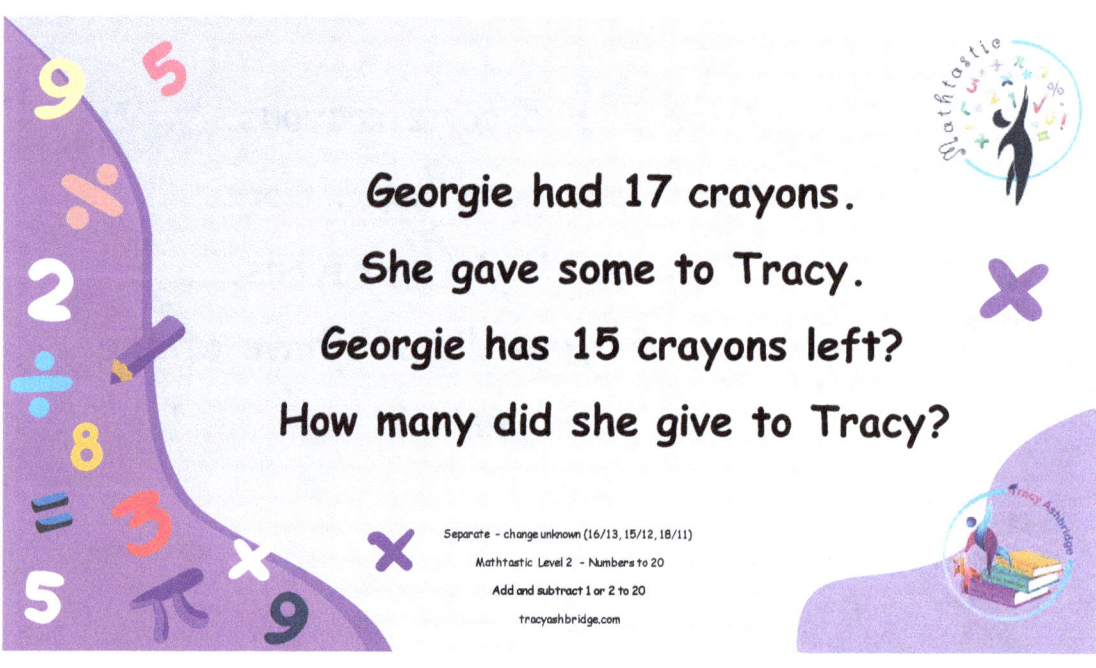

Georgie had 17 crayons.
She gave some to Tracy.
Georgie has 15 crayons left?
How many did she give to Tracy?

Separate – change unknown (16/13, 15/12, 18/11)
Mathtastic Level 2 – Numbers to 20
Add and subtract 1 or 2 to 20
tracyashbridge.com

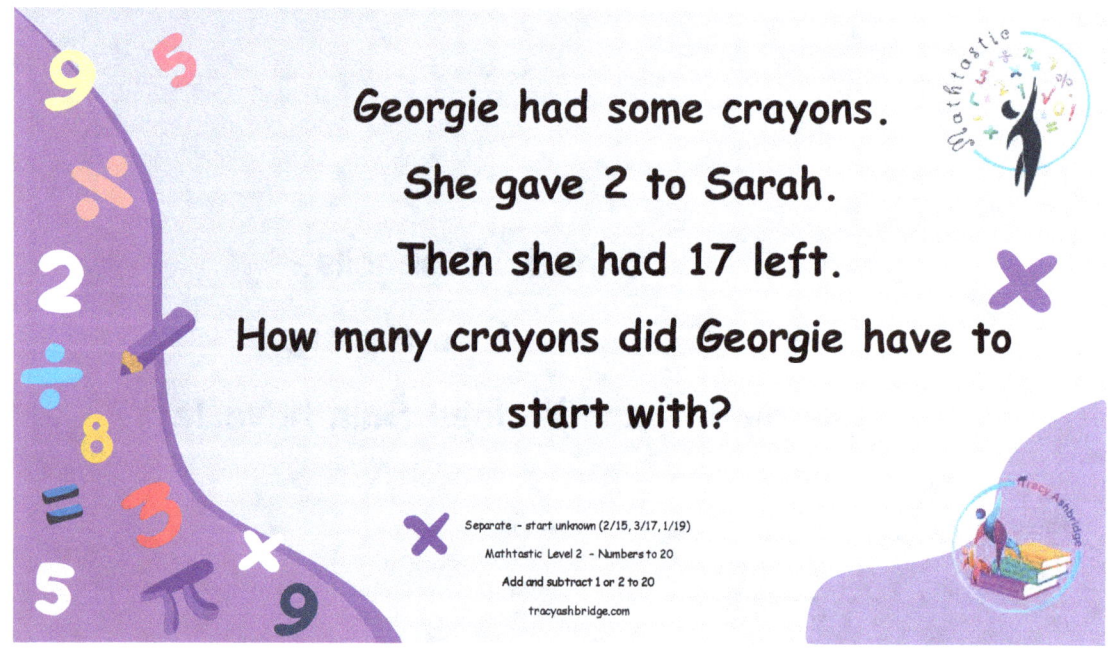

Georgie had some crayons.
She gave 2 to Sarah.
Then she had 17 left.
How many crayons did Georgie have to start with?

Separate – start unknown (2/15, 3/17, 1/19)
Mathtastic Level 2 – Numbers to 20
Add and subtract 1 or 2 to 20
tracyashbridge.com

14 green crayons and 2 blue crayons were in the pot.
How many were in the pot altogether?

Part - Part- Whole - whole unknown (17/3, 15/1, 12/3)
Mathtastic Level 2 - Numbers to 20
Add and subtract 1 or 2 to 20
tracyashbridge.com

There were 16 red and yellow crayons.
14 were red.
How many were yellow?

Part - Part- Whole - part unknown (15/12, 18/17, 16/13)
Mathtastic Level 2 - Numbers to 20
Add and subtract 1 or 2 to 20
tracyashbridge.com

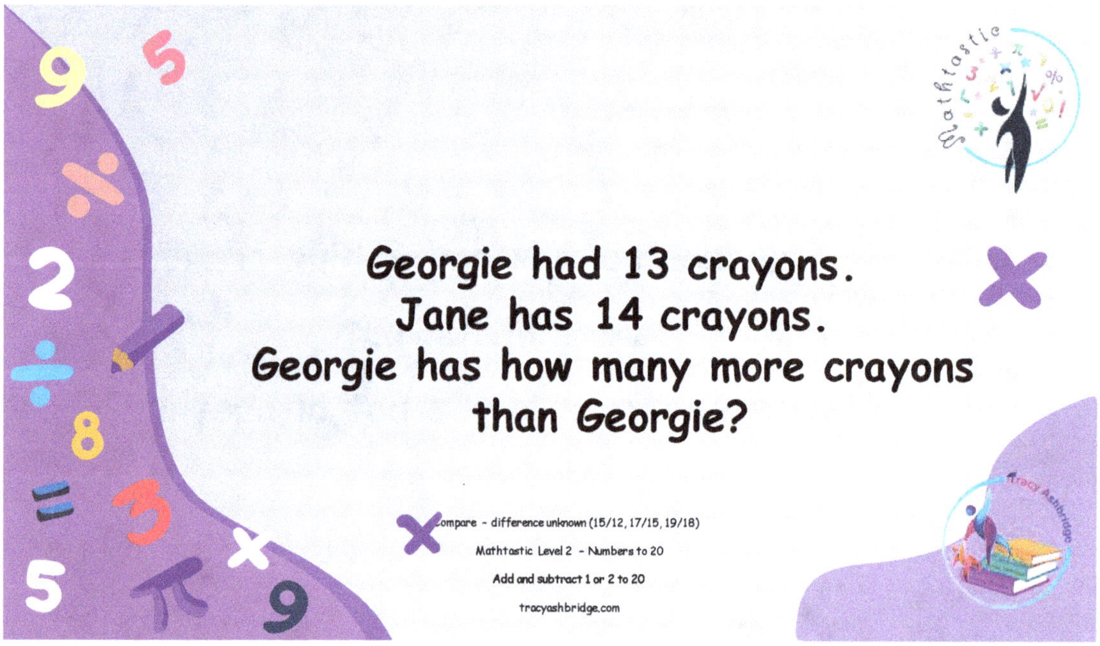
Mathtastic
Georgie had 13 crayons.
Jane has 14 crayons.
Georgie has how many more crayons than Georgie?
Compare - difference unknown (15/12, 17/15, 19/18)
Mathtastic Level 2 - Numbers to 20
Add and subtract 1 or 2 to 20
tracyashbridge.com
Tracy Ashbridge

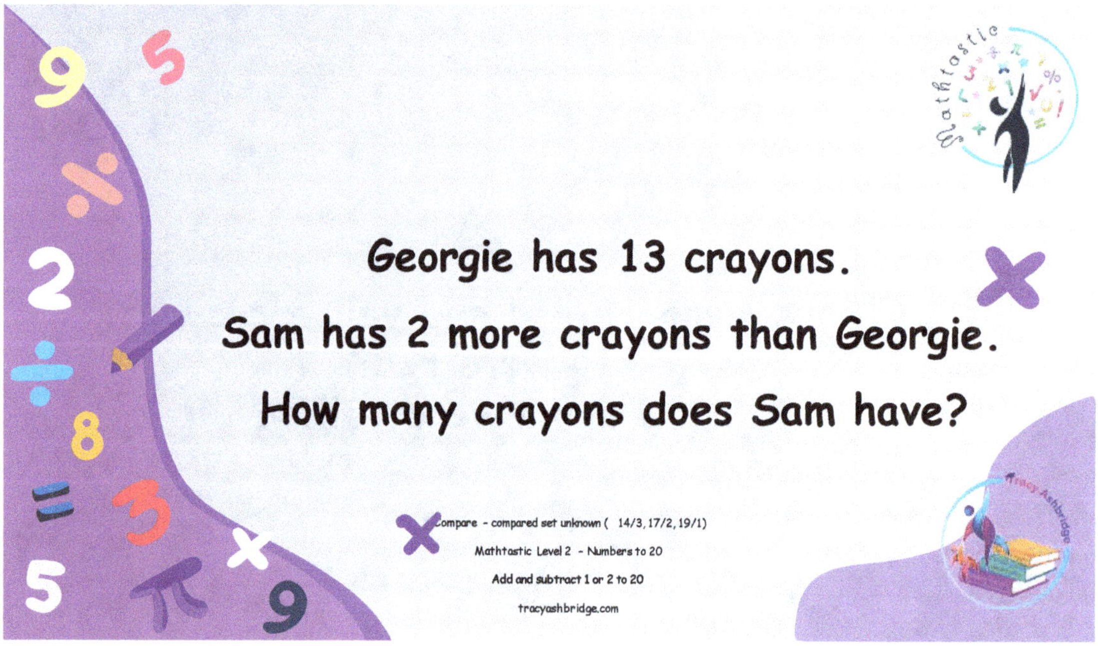
Mathtastic
Georgie has 13 crayons.
Sam has 2 more crayons than Georgie.
How many crayons does Sam have?
Compare - compared set unknown (14/3, 17/2, 19/1)
Mathtastic Level 2 - Numbers to 20
Add and subtract 1 or 2 to 20
tracyashbridge.com
Tracy Ashbridge

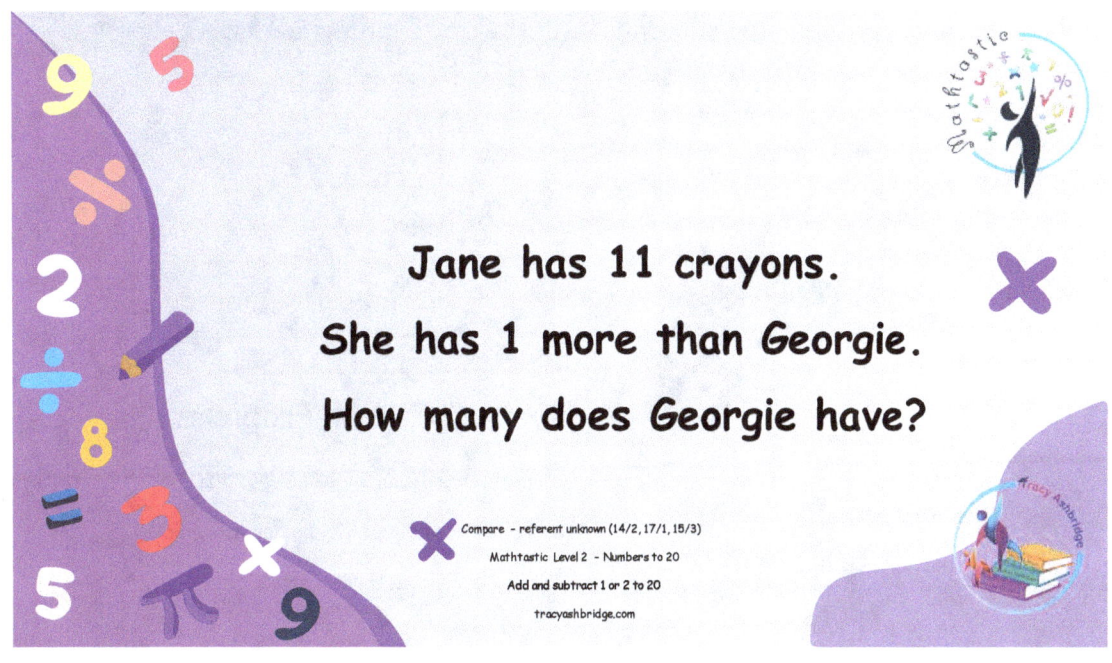

Jane has 11 crayons.

She has 1 more than Georgie.

How many does Georgie have?

Compare - referent unknown (14/2, 17/1, 15/3)

Mathtastic Level 2 - Numbers to 20

Add and subtract 1 or 2 to 20

tracyashbridge.com

Module 2 – Add from the largest number, subtract by counting back

Mathstastic

CGI Math Problems

Mathtastic Level 2 – Numbers to 20

Add from largest, subtract by counting back

tracyashbridge.com

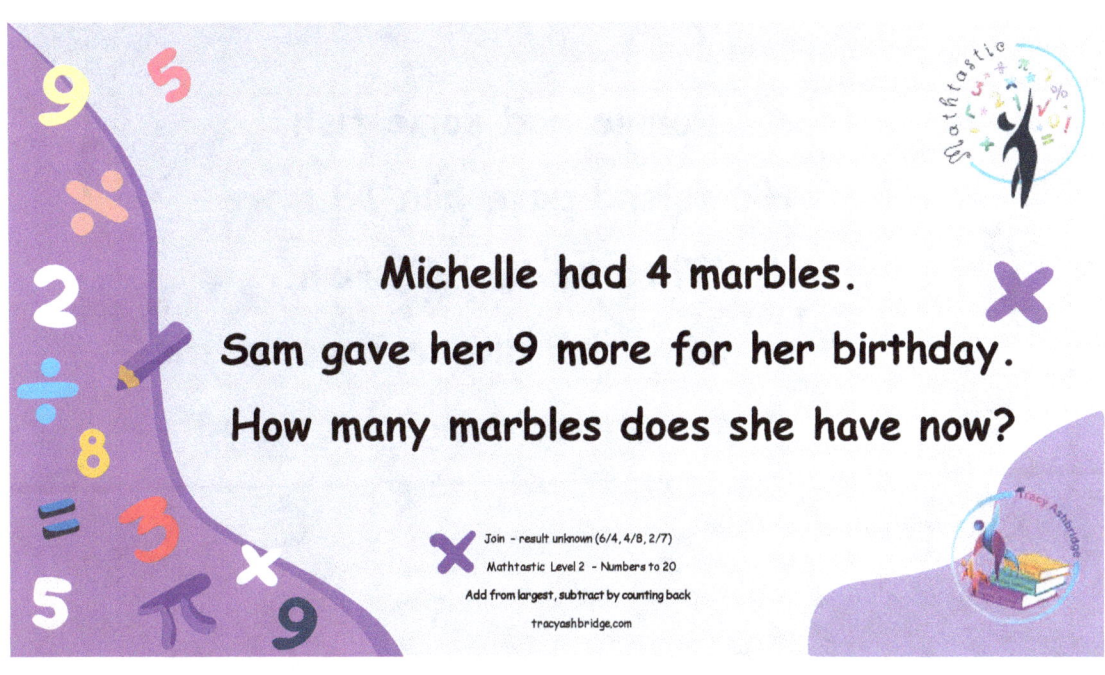

Michelle had 4 marbles.
Sam gave her 9 more for her birthday.
How many marbles does she have now?

Join – result unknown (6/4, 4/8, 2/7)
Mathtastic Level 2 – Numbers to 20
Add from largest, subtract by counting back
tracyashbridge.com

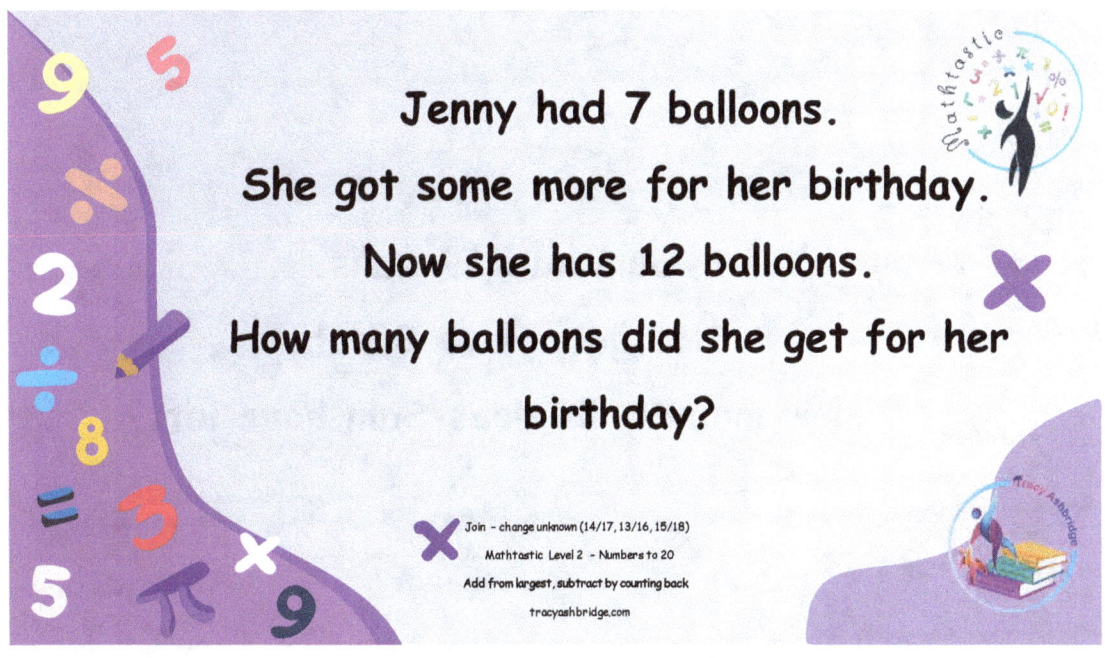

Jenny had 7 balloons.
She got some more for her birthday.
Now she has 12 balloons.
How many balloons did she get for her birthday?

Join – change unknown (14/17, 13/16, 15/18)
Mathtastic Level 2 – Numbers to 20
Add from largest, subtract by counting back
tracyashbridge.com

Ronnie had some fish.
His friend gave him 13 more.
Then he had 17 fish.
How many fish did he have at the beginning?

Join – start unknown (14/17, 9/11, 15/18)
Mathtastic Level 2 – Numbers to 20
Add from largest, subtract by counting back
tracyashbridge.com

Sam had 19 balls.
He gave 4 to Sarah.
How many balls does Sam have left?

Separate – result unknown (15/3, 17/4, 19/4)
Mathtastic Level 2 – Numbers to 20
Add from largest, subtract by counting back
tracyashbridge.com

Dave had 15 cards.
He gave some to Tracy.
Dave has 11 cards left?
How many did he give to Tracy?

Separate - change unknown (17/13, 18/16, 16/15)
Mathtastic Level 2 - Numbers to 20
Add from largest, subtract by counting back
tracyashbridge.com

Tim had some hats.
He gave 14 to Flynn.
Then he had 2 left.
How many hats did Tim have to start with?

Separate - start unknown (4/12, 3/16, 16/1)
Mathtastic Level 2 - Numbers to 20
Add from largest, subtract by counting back
tracyashbridge.com

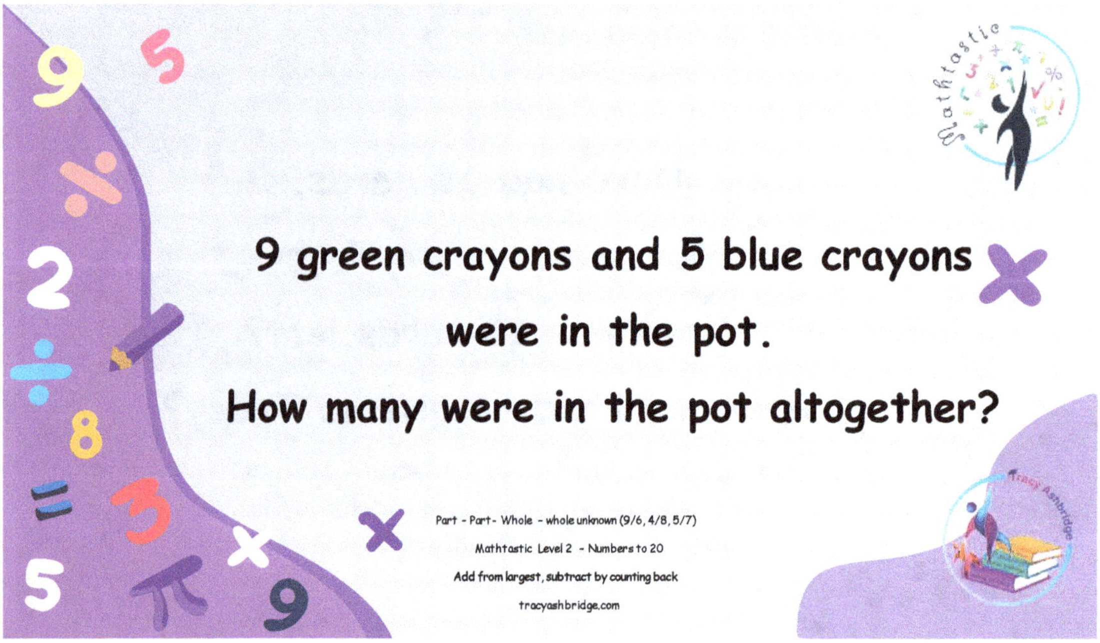

9 green crayons and 5 blue crayons were in the pot.
How many were in the pot altogether?

Part - Part - Whole - whole unknown (9/6, 4/8, 5/7)
Mathtastic Level 2 - Numbers to 20
Add from largest, subtract by counting back
tracyashbridge.com

There were 18 crayons.
4 were red.
How many were yellow?

Part - Part - Whole - part unknown (17/5, 14/12, 18/3, 19/16)
Mathtastic Level 2 - Numbers to 20
Add from largest, subtract by counting back
tracyashbridge.com

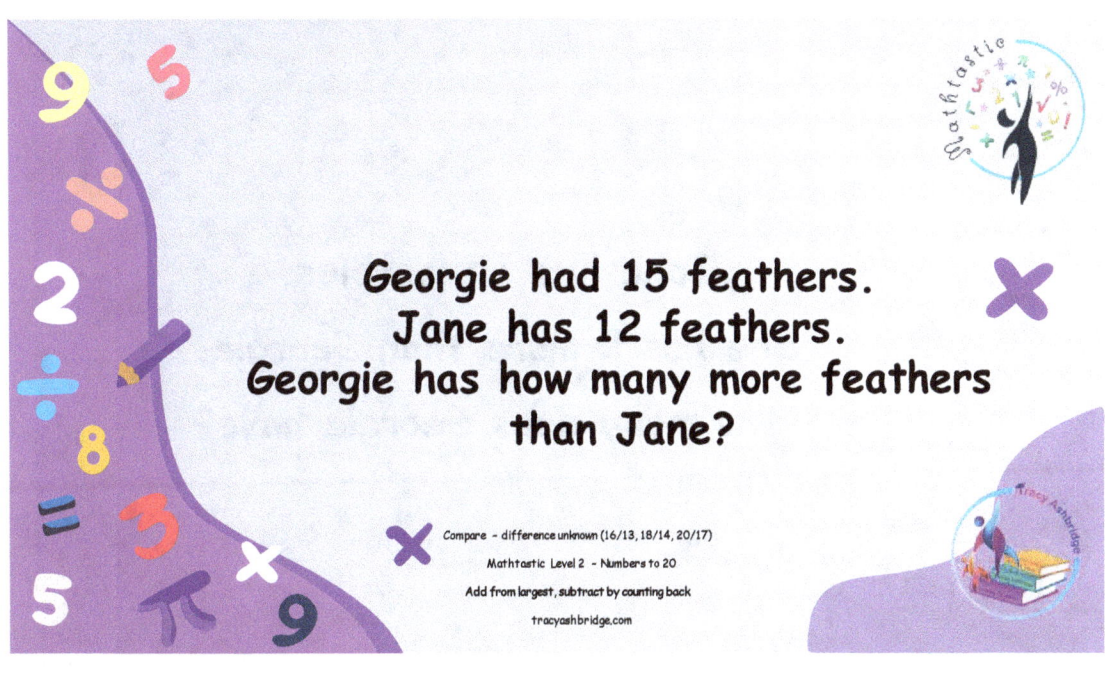

Georgie had 15 feathers.
Jane has 12 feathers.
Georgie has how many more feathers than Jane?

Compare - difference unknown (16/13, 18/14, 20/17)
Mathtastic Level 2 - Numbers to 20
Add from largest, subtract by counting back
tracyashbridge.com

Jason has 12 crabs.

Sam has 7 more crabs than Jason.

How many crabs does Sam have?

Compare - compared set unknown (13/5, 7/9, 5/11)
Mathtastic Level 2 - Numbers to 20
Add from largest, subtract by counting back
tracyashbridge.com

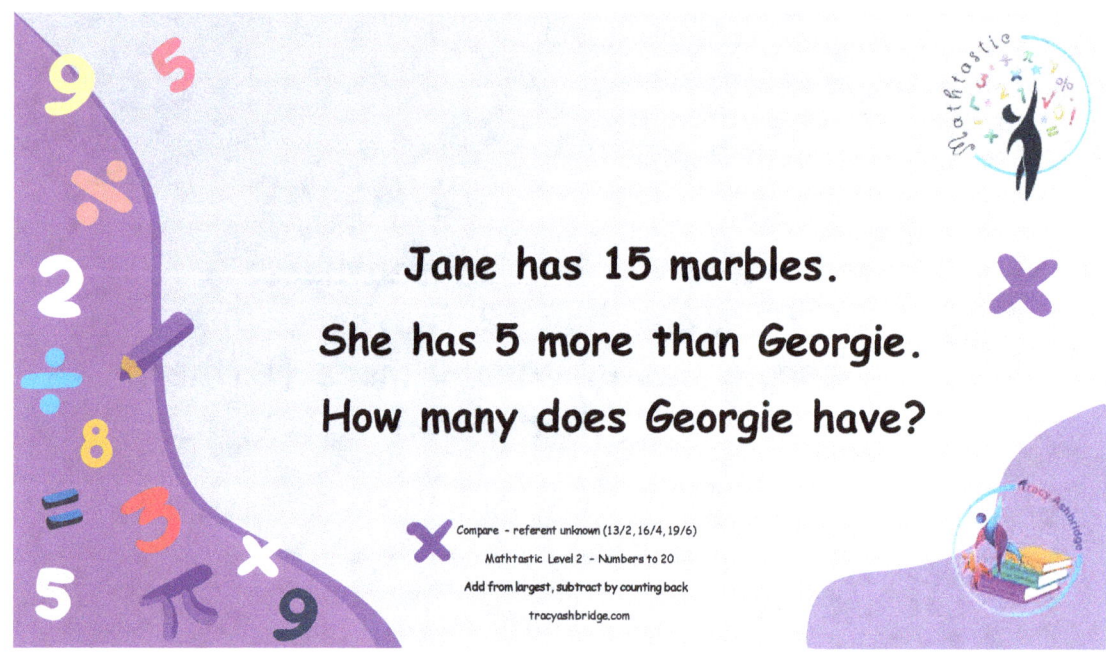

Jane has 15 marbles.

She has 5 more than Georgie.

How many does Georgie have?

Compare – referent unknown (13/2, 16/4, 19/6)
Mathtastic Level 2 – Numbers to 20
Add from largest, subtract by counting back
tracyashbridge.com

Module 3 – Rainbow facts, pairs of numbers to 20

Michelle had 19 crayons.
Sam gave her 1 more for her birthday.
How many crayons does she have now?

Join – result unknown (14/6, 17/3, 18/2)
Mathtastic Level 2 – Numbers to 20
Rainbow Facts
tracyashbridge.com

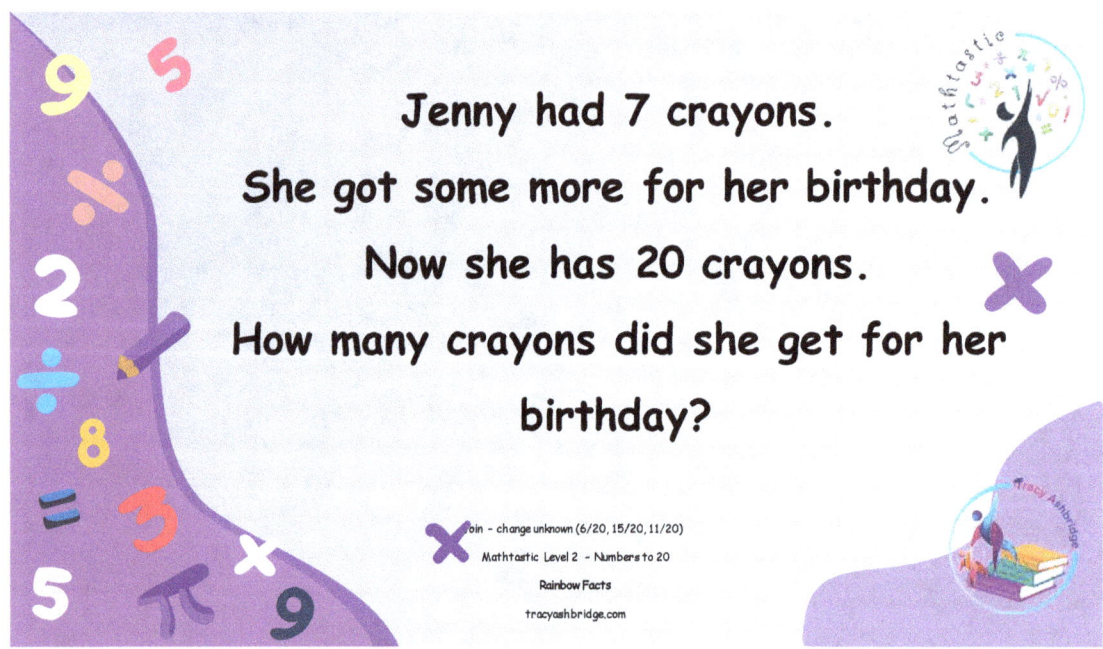

Jenny had 7 crayons.
She got some more for her birthday.
Now she has 20 crayons.
How many crayons did she get for her birthday?

Join – change unknown (6/20, 15/20, 11/20)
Mathtastic Level 2 – Numbers to 20
Rainbow Facts
tracyashbridge.com

Ron had some fish.
His friend gave him 7 more.
Then he had 13 fish.
How many fish did he have at the beginning?

Join - start unknown (6/11, 7/15, 8/15)
Mathtastic Level 2 - Numbers to 20
Rainbow Facts
tracyashbridge.com

Sam had 20 balls.
He gave 9 to Sarah.
How many balls does Sam have left?

Separate - result unknown (20/14, 20/3, 20/5)
Mathtastic Level 2 - Numbers to 20
Rainbow Facts
tracyashbridge.com

Dave had 20 cards.
He gave some to Tracy.
Dave has 5 cards left?
How many did he give to Tracy?

Separate – change unknown (20/14, 20/17, 20/8)
Mathtastic Level 2 – Numbers to 20
Rainbow Facts
tracyashbridge.com

Tim had some hats.
He gave 4 to Flynn.
Then he had 16 left.
How many hats did Tim have to start with?

Separate – start unknown (5/15, 17/3, 2/18)
Mathtastic Level 2 – Numbers to 20
Rainbow Facts
tracyashbridge.com

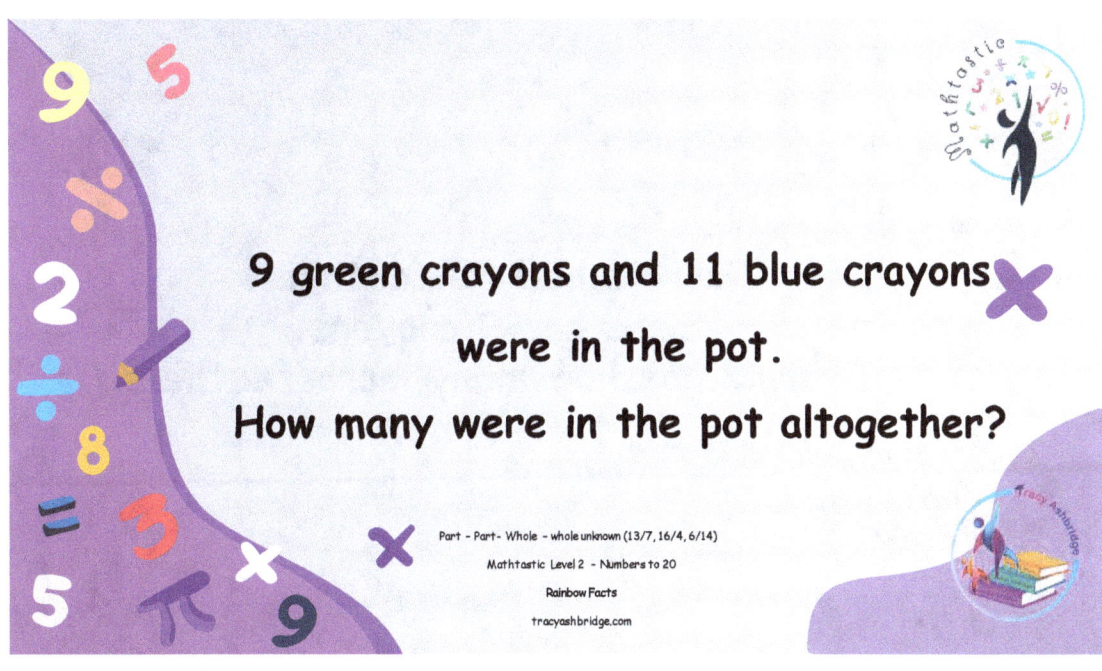

9 green crayons and 11 blue crayons were in the pot.
How many were in the pot altogether?

Part - Part - Whole - whole unknown (13/7, 16/4, 6/14)
Mathtastic Level 2 - Numbers to 20
Rainbow Facts
tracyashbridge.com

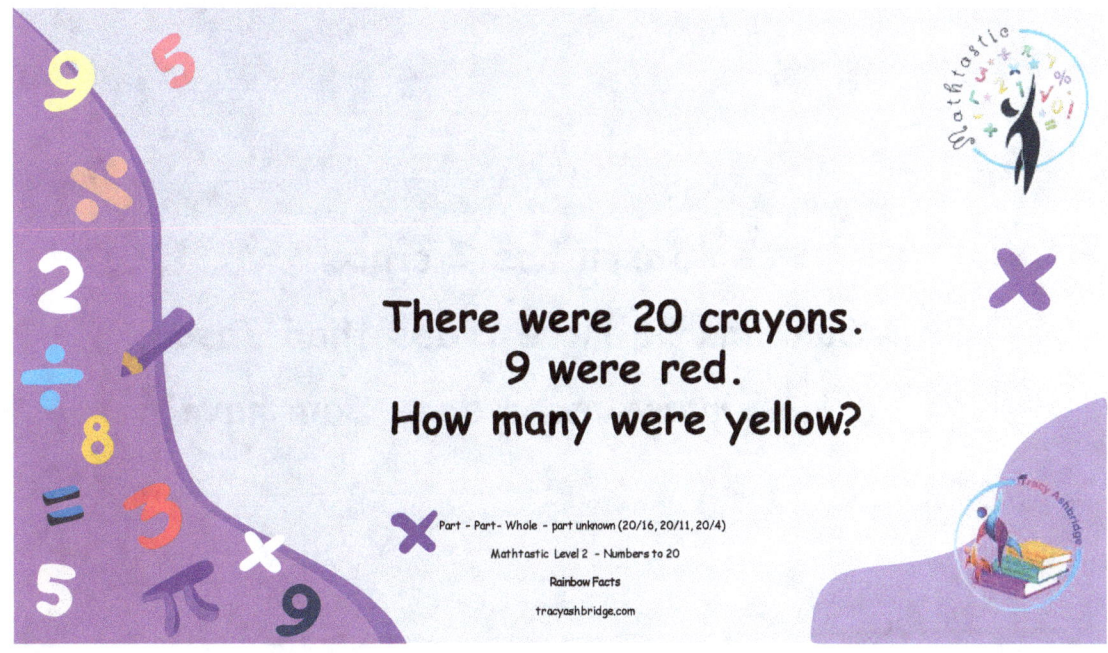

There were 20 crayons.
9 were red.
How many were yellow?

Part - Part - Whole - part unknown (20/16, 20/11, 20/4)
Mathtastic Level 2 - Numbers to 20
Rainbow Facts
tracyashbridge.com

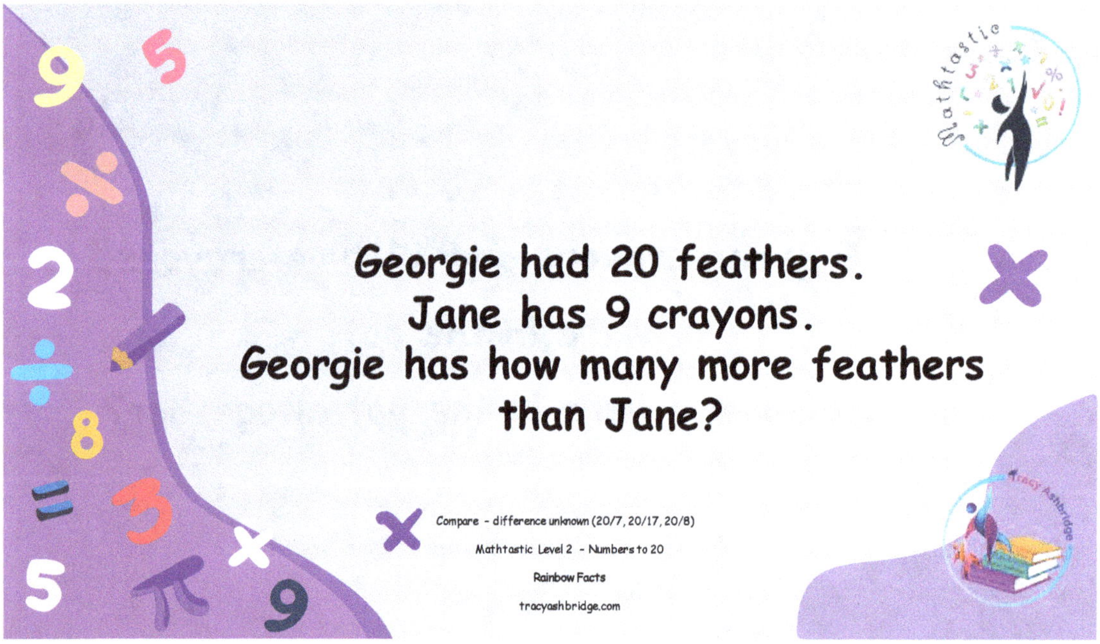

Georgie had 20 feathers.
Jane has 9 crayons.
Georgie has how many more feathers than Jane?

Compare - difference unknown (20/7, 20/17, 20/8)

Mathtastic Level 2 - Numbers to 20

Rainbow Facts

tracyashbridge.com

Jason has 7 crabs.

Sam has 13 more crabs than Jason.

How many crabs does Sam have?

Compare - compared set unknown (12/8, 14/6, 2/18)

Mathtastic Level 2 - Numbers to 20

Rainbow Facts

tracyashbridge.com

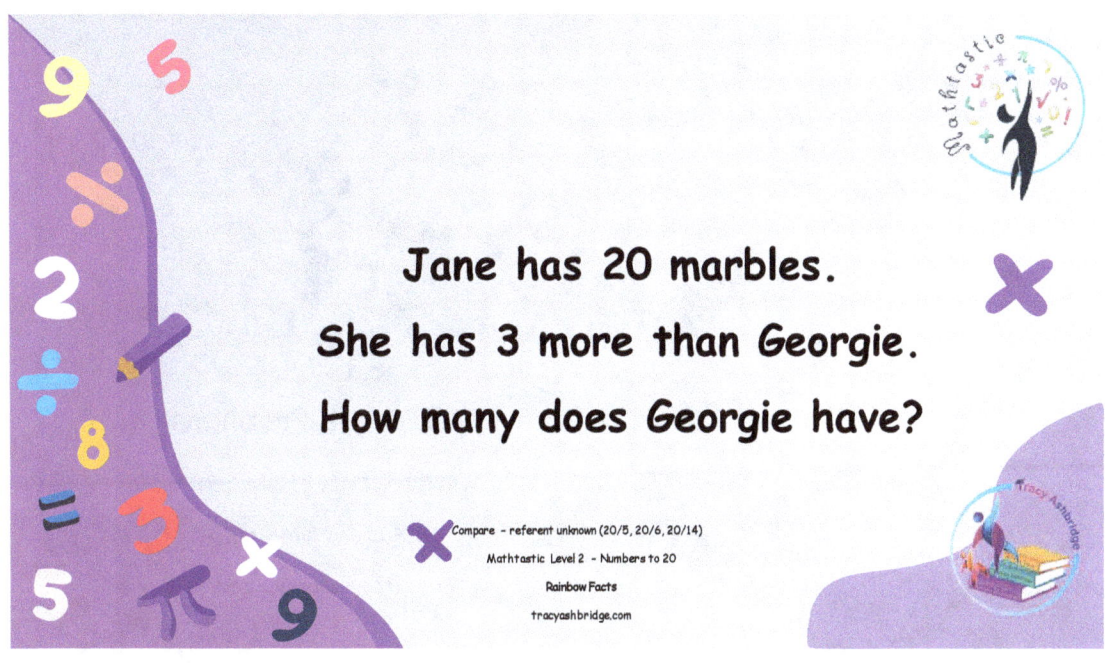

Jane has 20 marbles.
She has 3 more than Georgie.
How many does Georgie have?

Compare – referent unknown (20/5, 20/6, 20/14)
Mathtastic Level 2 – Numbers to 20
Rainbow Facts
tracyashbridge.com

Module 4- Add and subtract 10

Mathstastic
CGI Math Problems

Mathtastic Level 2 – Numbers to 20

Add and subtract 10

tracyashbridge.com

Michelle had 10 marbles.
Sam gave her 9 more for her birthday.
How many marbles does she have now?

Join - result unknown (10/4, 6/10, 7/10)
Mathtastic Level 2 - Numbers to 20
Add and subtract 10
tracyashbridge.com

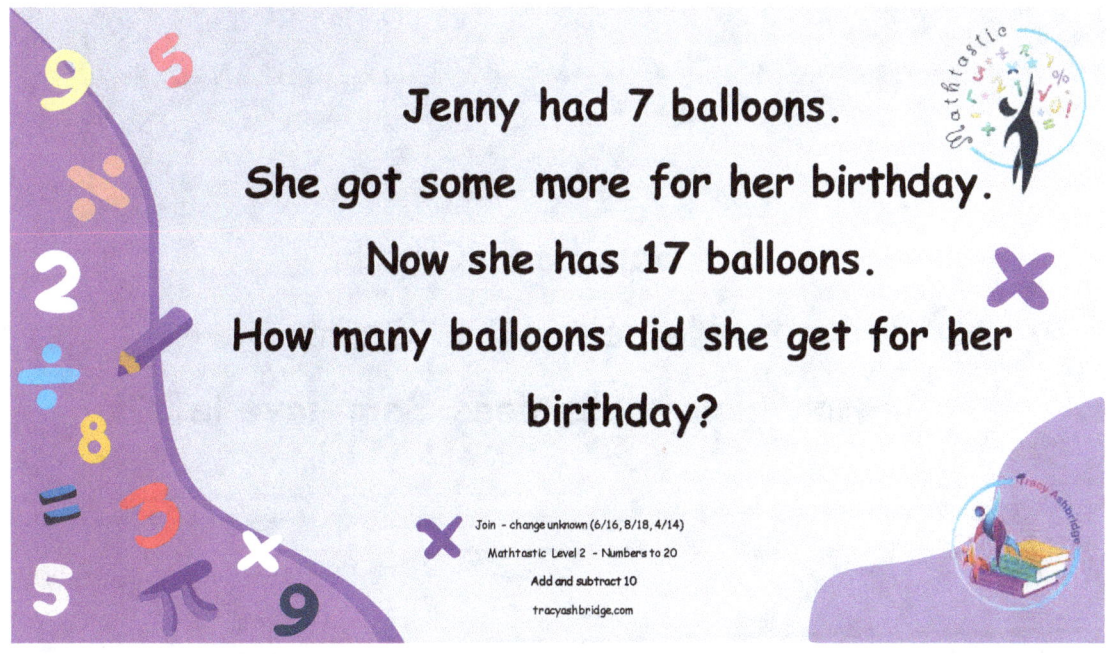

Jenny had 7 balloons.
She got some more for her birthday.
Now she has 17 balloons.
How many balloons did she get for her birthday?

Join - change unknown (6/16, 8/18, 4/14)
Mathtastic Level 2 - Numbers to 20
Add and subtract 10
tracyashbridge.com

Ronnie had some fish.
His friend gave him 7 more.
Then he had 17 fish.
How many fish did he have at the beginning?

Join - start unknown (10/13, 6/16, 9/19)
Mathtastic Level 2 - Numbers to 20
Add and subtract 10
tracyashbridge.com

Sam had 20 balls.
He gave 10 to Sarah.
How many balls does Sam have left?

Separate - result unknown (15/5, 18/10, 12/10)
Mathtastic Level 2 - Numbers to 20
Add and subtract 10
tracyashbridge.com

Dave had 15 cards.
He gave some to Tracy.
Dave has 5 cards left?
How many did he give to Tracy?

Separate – change unknown (17/10, 13/3, 18/10)
Mathtastic Level 2 – Numbers to 20
Add and subtract 10
tracyashbridge.com

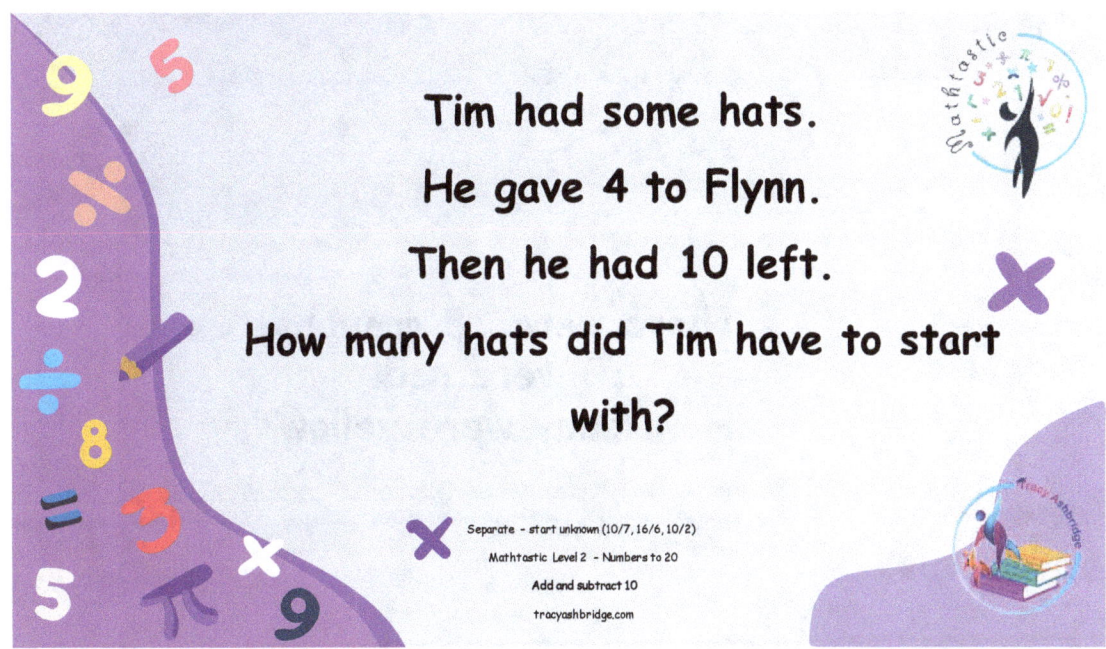

Tim had some hats.
He gave 4 to Flynn.
Then he had 10 left.
How many hats did Tim have to start with?

Separate – start unknown (10/7, 16/6, 10/2)
Mathtastic Level 2 – Numbers to 20
Add and subtract 10
tracyashbridge.com

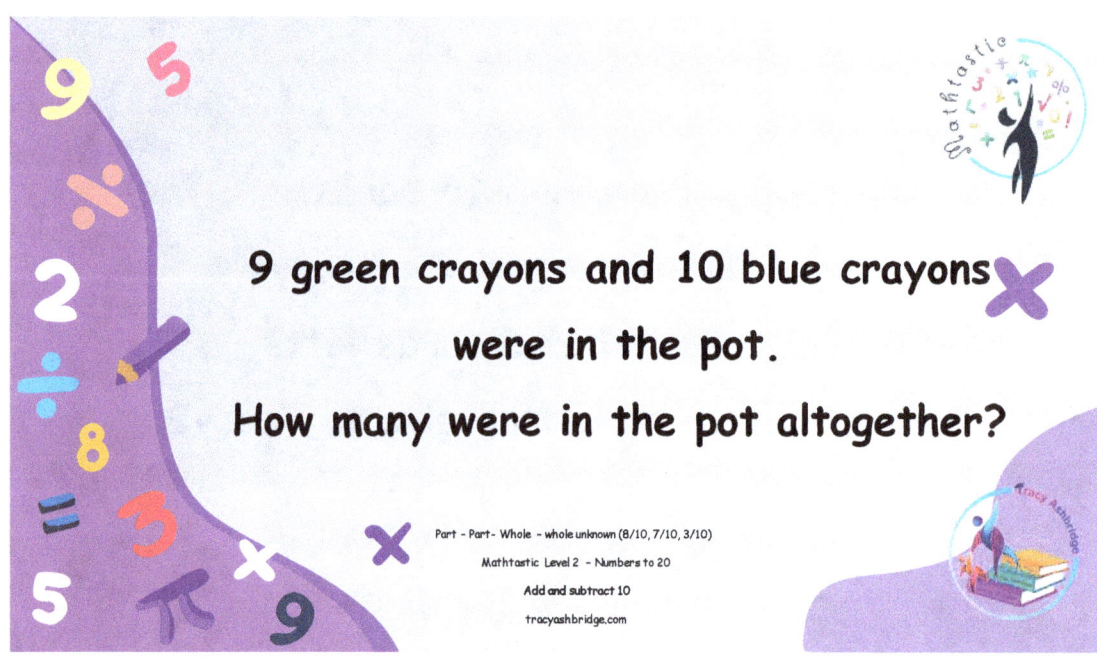

9 green crayons and 10 blue crayons were in the pot.
How many were in the pot altogether?

Part – Part – Whole – whole unknown (8/10, 7/10, 3/10)
Mathtastic Level 2 – Numbers to 20
Add and subtract 10
tracyashbridge.com

There were 18 crayons.
10 were red.
How many were yellow?

Part – Part – Whole – part unknown (17/7, 16/10, 14/4)
Mathtastic Level 2 – Numbers to 20
Add and subtract 10
tracyashbridge.com

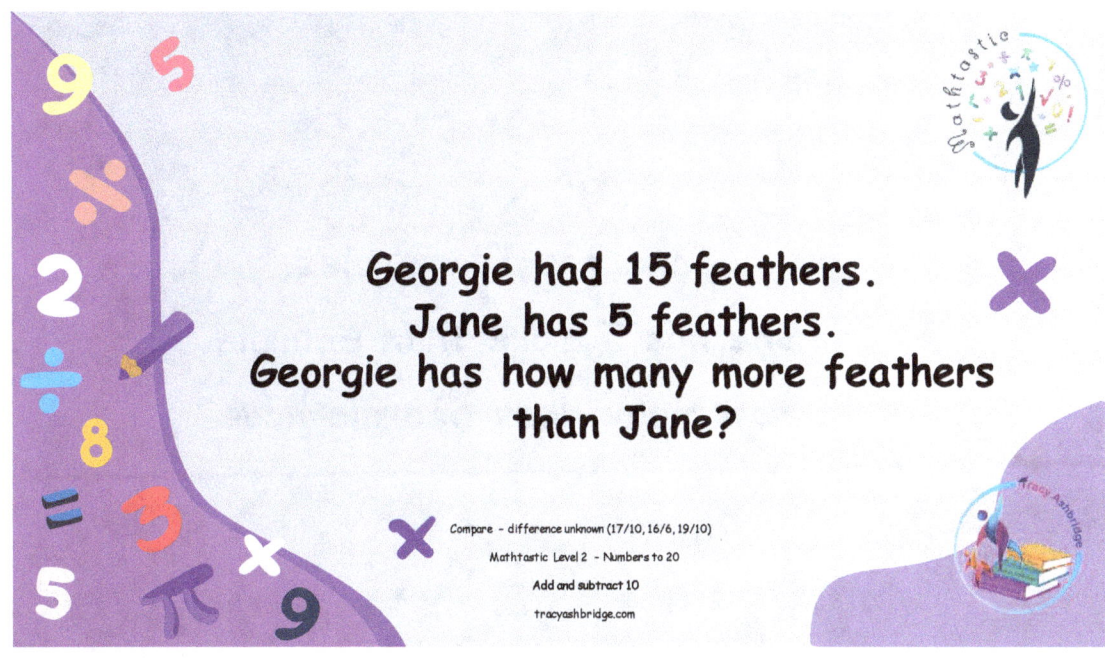

Georgie had 15 feathers.
Jane has 5 feathers.
Georgie has how many more feathers than Jane?

Compare - difference unknown (17/10, 16/6, 19/10)
Mathtastic Level 2 - Numbers to 20
Add and subtract 10
tracyashbridge.com

Jason has 7 crabs.

Sam has 10 more crabs than Jason.

How many crabs does Sam have?

Compare - compared set unknown (6/10, 10/9, 3/10)
Mathtastic Level 2 - Numbers to 20
Add and subtract 10
tracyashbridge.com

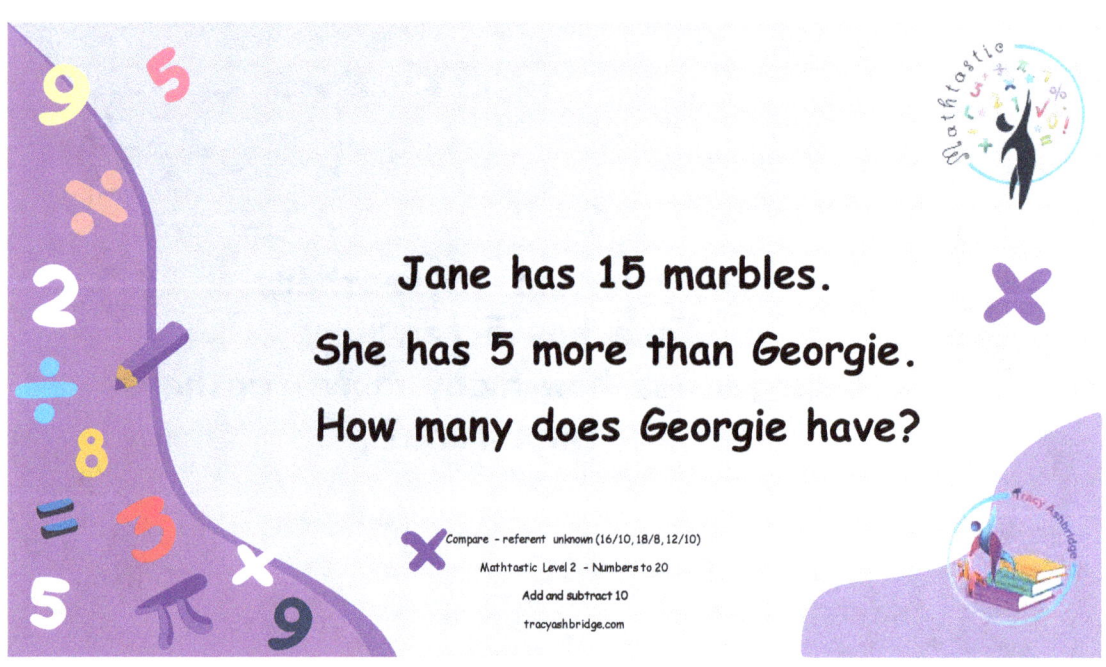

Jane has 15 marbles.

She has 5 more than Georgie.

How many does Georgie have?

Compare - referent unknown (16/10, 18/8, 12/10)
Mathtastic Level 2 - Numbers to 20
Add and subtract 10
tracyashbridge.com

Module 5- Doubles to 20

John had 7 crayons.
Sam gave him 7 more for his birthday.
How many crayons does he have now?

Join - result unknown (10/10, 8/8, 6/6)
Mathtastic Level 2 - Numbers to 20
Doubles to 20
tracyashbridge.com

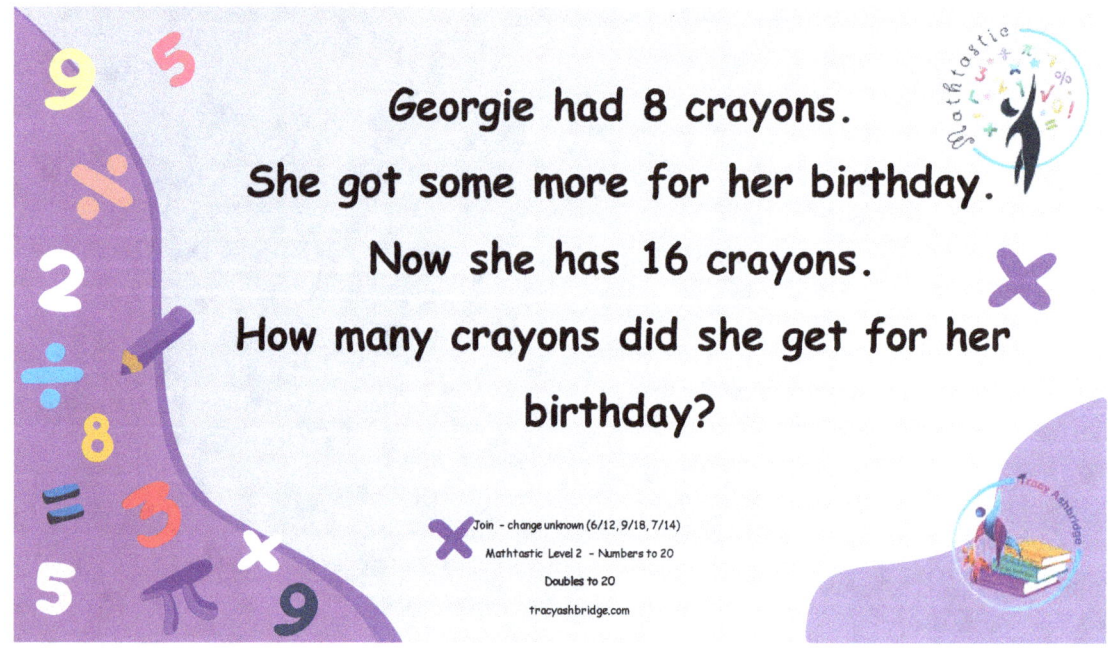

Georgie had 8 crayons.
She got some more for her birthday.
Now she has 16 crayons.
How many crayons did she get for her birthday?

Join - change unknown (6/12, 9/18, 7/14)
Mathtastic Level 2 - Numbers to 20
Doubles to 20
tracyashbridge.com

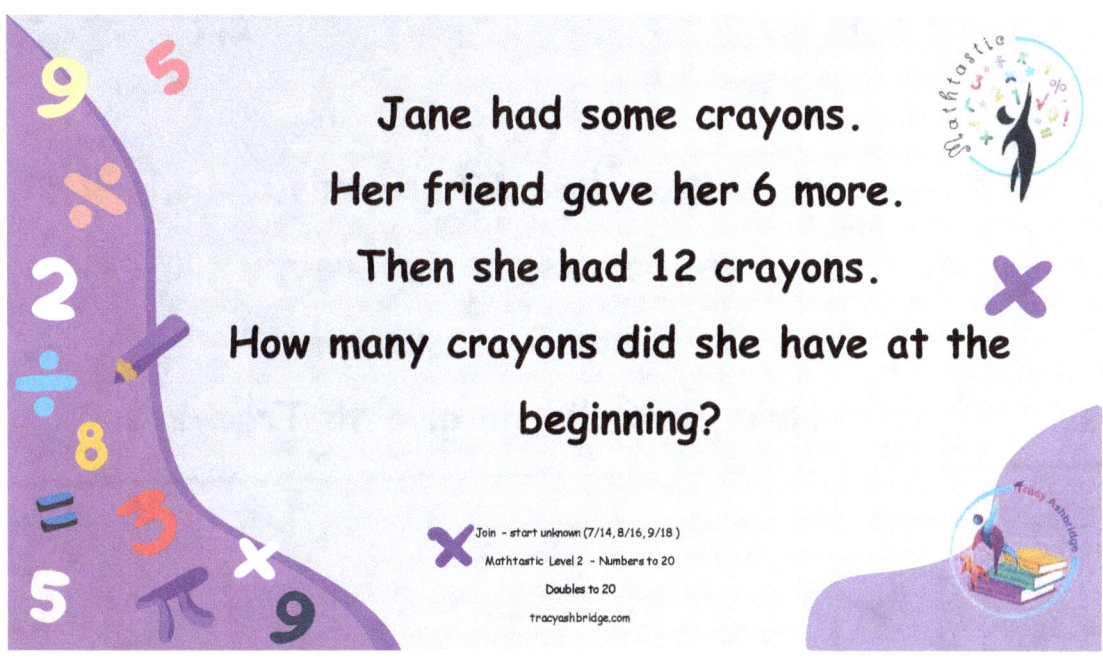

Jane had some crayons.
Her friend gave her 6 more.
Then she had 12 crayons.
How many crayons did she have at the beginning?

Join – start unknown (7/14, 8/16, 9/18)
Mathtastic Level 2 – Numbers to 20
Doubles to 20
tracyashbridge.com

Sam had 16 pencils.
He gave 8 to Sarah.
How many pencils does Sam have left?

Separate – result unknown (14/7, 18/9, 16/8)
Mathtastic Level 2 – Numbers to 20
Doubles to 20
tracyashbridge.com

Dave had 18 crayons.
He gave some to Tracy.
Dave has 9 crayons left?
How many did he give to Tracy?

Separate - change unknown (12/6, 18/9, 14/7)
Mathtastic Level 2 - Numbers to 20
Doubles to 20
tracyashbridge.com

Tim had some crayons.
He gave 6 to Flynn.
Then he had 6 left.
How many crayons did Tim have to start with?

Separate - start unknown (7/7, 9/9, 6/6)
Mathtastic Level 2 - Numbers to 20
Doubles to 20
tracyashbridge.com

9 green crayons and 9 blue crayons were in the pot.
How many were in the pot altogether?

Part – Part – Whole – whole unknown (6/6, 8/8, 9/9)
Mathtastic Level 2 – Numbers to 20
Doubles to 20
tracyashbridge.com

There were 20 red and yellow crayons.
10 were red.
How many were yellow?

Part – Part – Whole – part unknown (14/7, 16/8, 12/6)
Mathtastic Level 2 – Numbers to 20
Doubles to 20
tracyashbridge.com

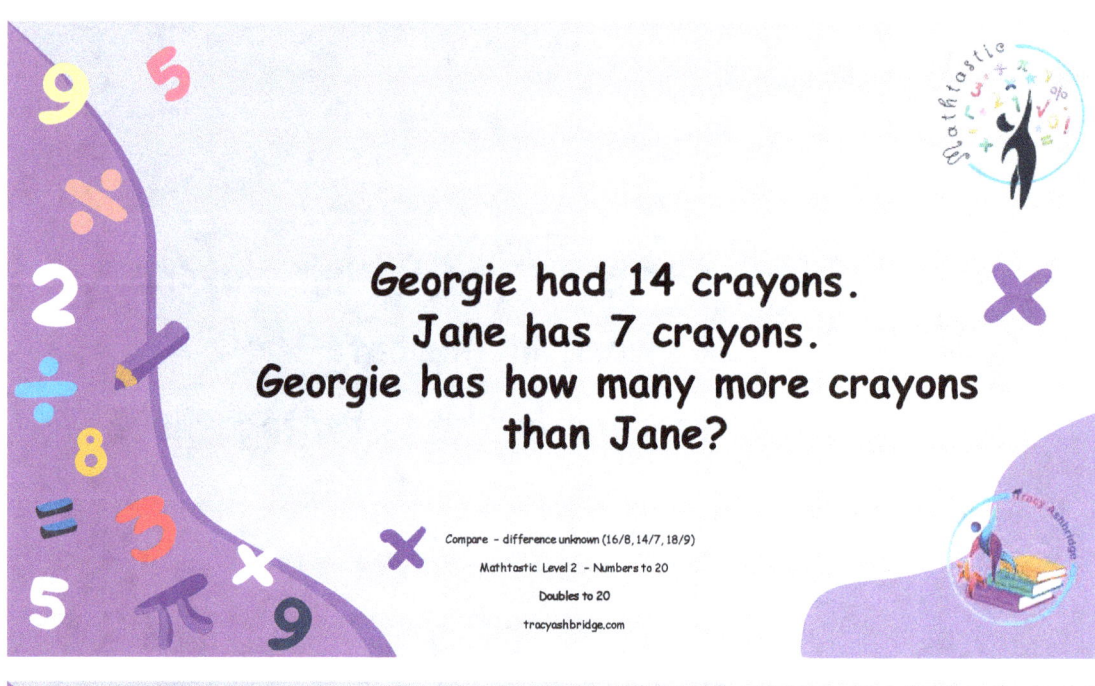

Georgie had 14 crayons.
Jane has 7 crayons.
Georgie has how many more crayons than Jane?

Compare - difference unknown (16/8, 14/7, 18/9)
Mathtastic Level 2 - Numbers to 20
Doubles to 20
tracyashbridge.com

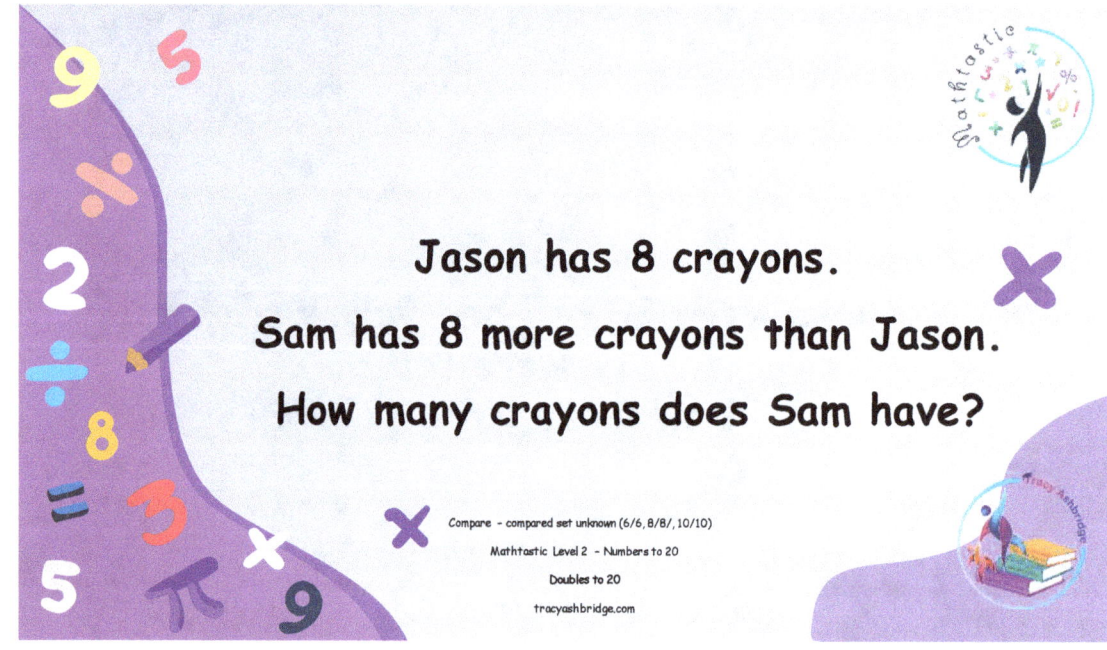

Jason has 8 crayons.

Sam has 8 more crayons than Jason.

How many crayons does Sam have?

Compare - compared set unknown (6/6, 8/8/, 10/10)
Mathtastic Level 2 - Numbers to 20
Doubles to 20
tracyashbridge.com

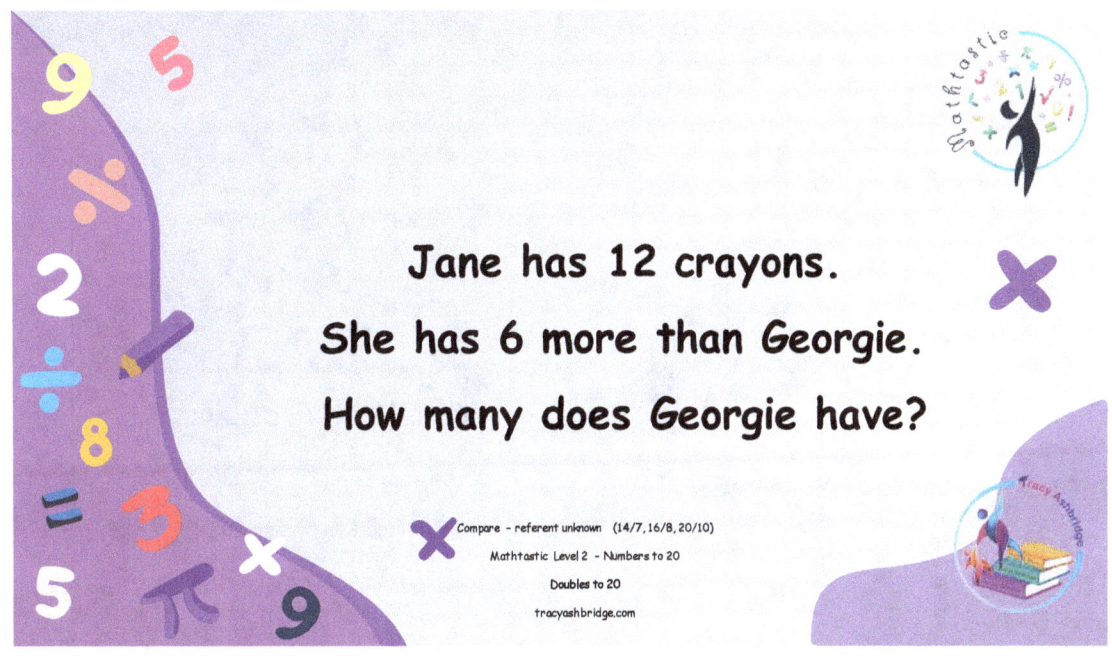

Jane has 12 crayons.
She has 6 more than Georgie.
How many does Georgie have?

Compare - referent unknown   (14/7, 16/8, 20/10)
Mathtastic Level 2 - Numbers to 20
Doubles to 20
tracyashbridge.com

Module 6 – Near doubles

Jack had 9 crayons.
Sam gave him 10 more for his birthday.
How many crayons does he have now?

Join – result unknown (7/8, 6/7, 9/8)
Mathtastic Level 2 – Numbers to 20
Near doubles to 20
tracyashbridge.com

Georgie had 7 crayons.
She got some more for her birthday.
Now she has 15 crayons.
How many crayons did she get for her birthday?

Join – change unknown (8/17, 6/13, 7,13)
Mathtastic Level 2 – Numbers to 20
Near doubles to 20
tracyashbridge.com

Jane had some crayons.
Her friend gave her 7 more.
Then she had 13 crayons.
How many crayons did she have at the beginning?

Join – start unknown (6/11, 7/15, 8/15)
Mathtastic Level 2 – Numbers to 20
Near doubles to 20
tracyashbridge.com

Sam had 17 pencils.
He gave 9 to Sarah.
How many pencils does Sam have left?

Separate – result unknown (13/7, 13/6, 19/10)
Mathtastic Level 2 – Numbers to 20
Near doubles to 20
tracyashbridge.com

Dave had 11 crayons.
He gave some to Tracy.
Dave has 5 crayons left?
How many did he give to Tracy?

Separate – change unknown (15/7, 17/8, 19/9)
Mathtastic Level 2 – Numbers to 20
Near doubles to 20
tracyashbridge.com

Tim had some crayons.
He gave 6 to Flynn.
Then he had 7 left.
How many crayons did Tim have to start with?

Separate – start unknown (7/8, 8/9, 5/6)
Mathtastic Level 2 – Numbers to 20
Near doubles to 20
tracyashbridge.com

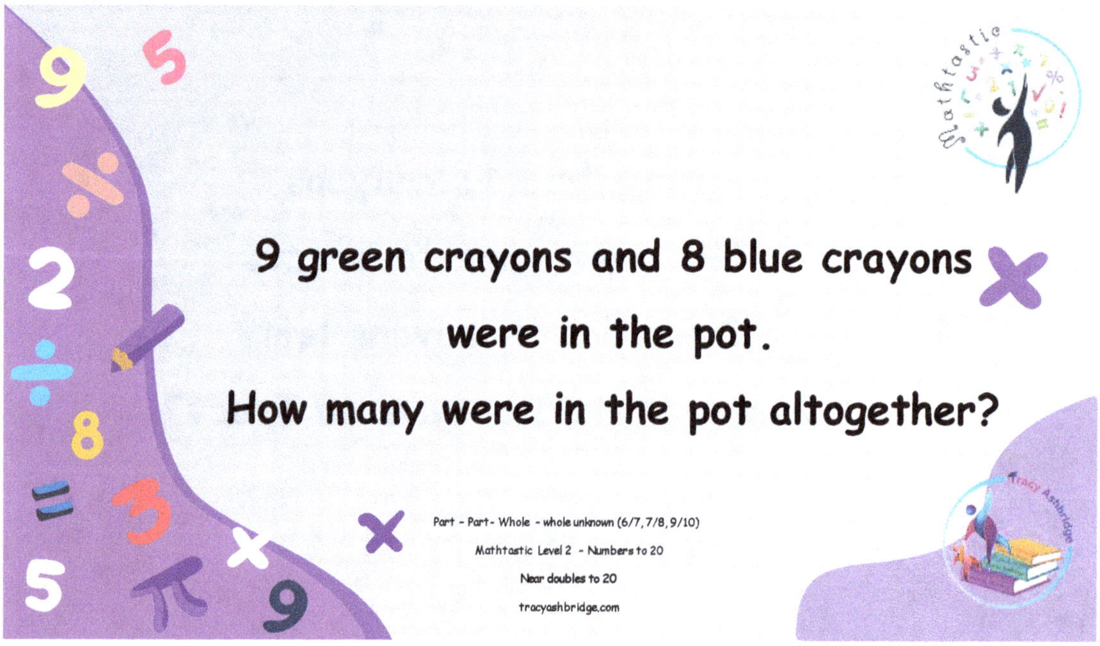

Mathtastic
9 green crayons and 8 blue crayons
were in the pot.
How many were in the pot altogether?
Part - Part- Whole - whole unknown (6/7, 7/8, 9/10)
Mathtastic Level 2 - Numbers to 20
Near doubles to 20
tracyashbridge.com
Tracy Ashbridge

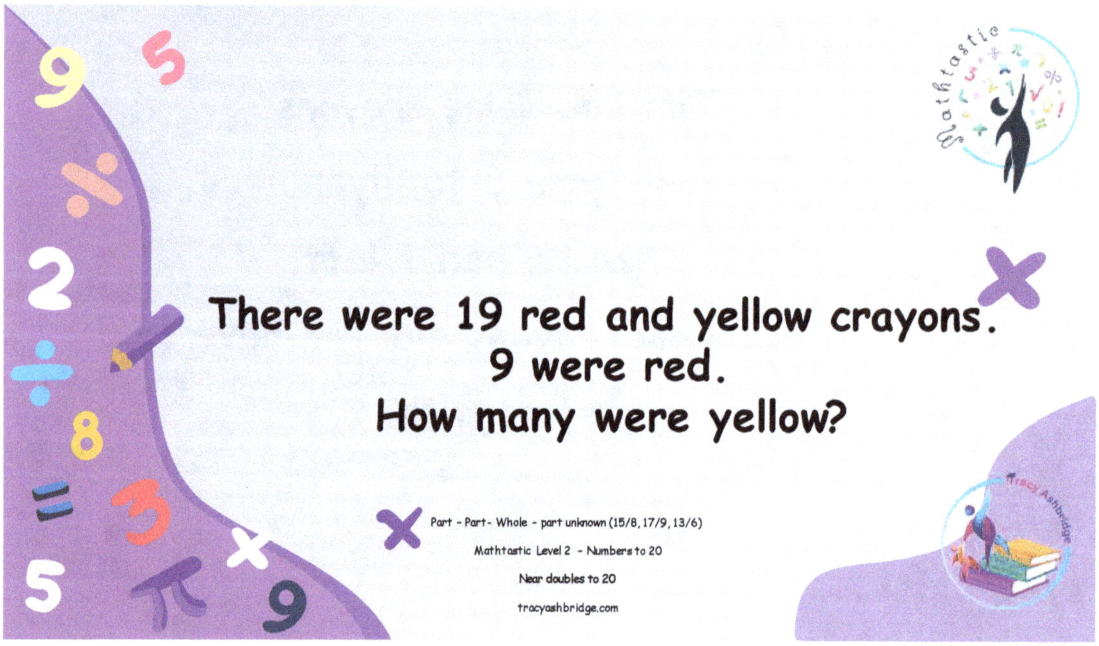

Mathtastic
There were 19 red and yellow crayons.
9 were red.
How many were yellow?
Part - Part- Whole - part unknown (15/8, 17/9, 13/6)
Mathtastic Level 2 - Numbers to 20
Near doubles to 20
tracyashbridge.com
Tracy Ashbridge

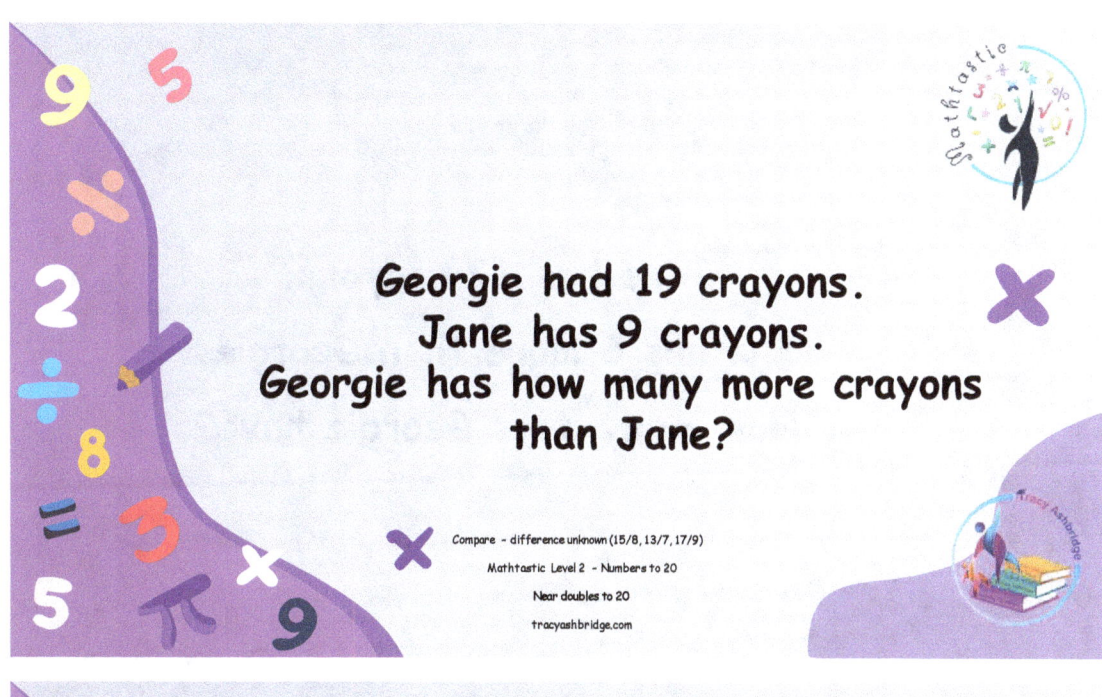

Georgie had 19 crayons.
Jane has 9 crayons.
Georgie has how many more crayons than Jane?

Compare - difference unknown (15/8, 13/7, 17/9)
Mathtastic Level 2 - Numbers to 20
Near doubles to 20
tracyashbridge.com

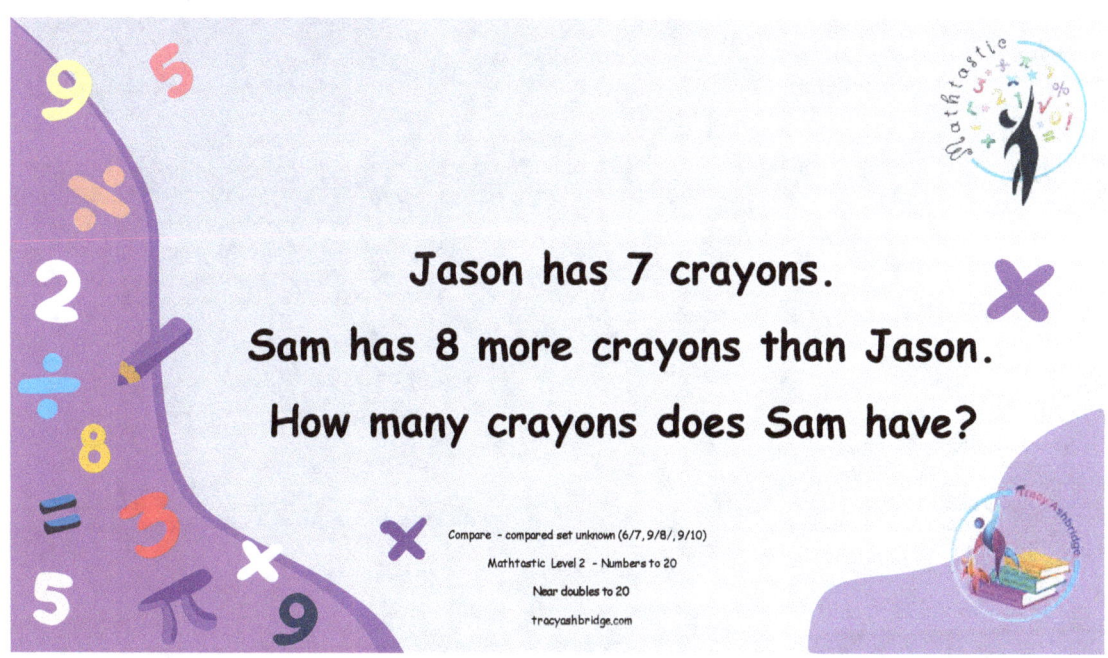

Jason has 7 crayons.

Sam has 8 more crayons than Jason.

How many crayons does Sam have?

Compare - compared set unknown (6/7, 9/8/, 9/10)
Mathtastic Level 2 - Numbers to 20
Near doubles to 20
tracyashbridge.com

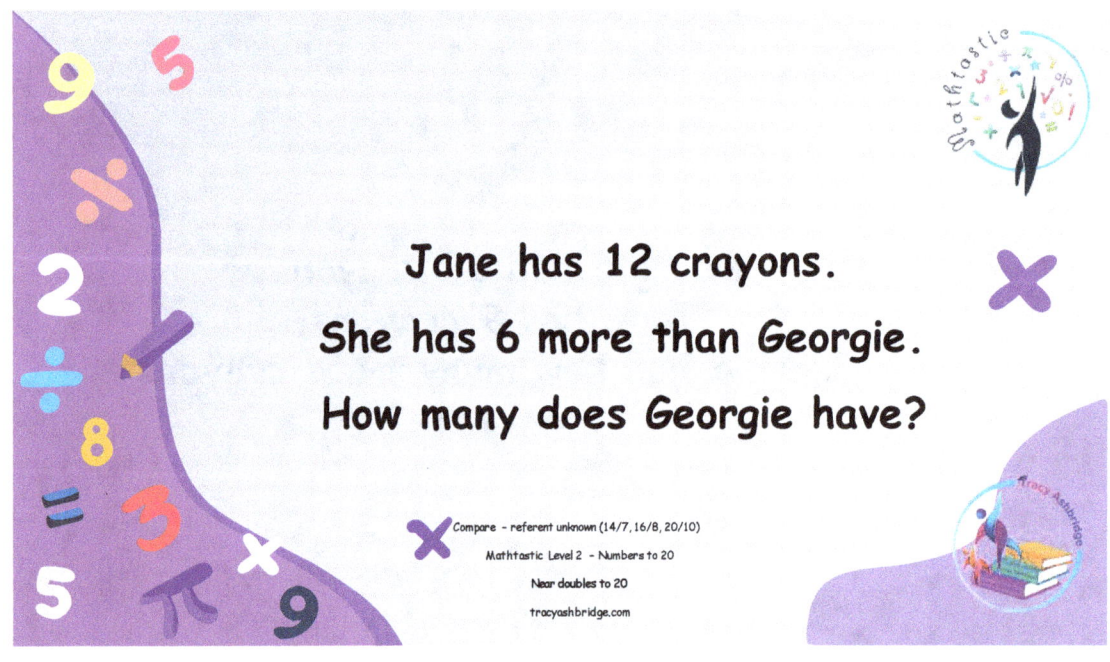

Jane has 12 crayons.
She has 6 more than Georgie.
How many does Georgie have?

Compare - referent unknown (14/7, 16/8, 20/10)
Mathtastic Level 2 - Numbers to 20
Near doubles to 20
tracyashbridge.com

Module 7 – Add using place value

This module overlaps with module 4 at this level. Please review missing pages from previous modules for revision of strategies

Module 8 – Add and subtract by compensation strategies

Michelle had 8 flowers.
Sam gave her 3 more for her birthday.
How many flowers does she have now?

Join – result unknown (9/4, 7/5, 11/6)
Mathtastic Level 2 – Numbers to 20
Add and Subtract by Compensation
tracyashbridge.com

Jenny had 6 candles.
She got some more for her birthday.
Now she has 11 candles.
How many candles did she get for her birthday?

Join – change unknown (8/12, 9/13, 14/17)
Mathtastic Level 2 – Numbers to 20
Add and Subtract by Compensation
tracyashbridge.com

Ronnie had some caterpillars.
His friend gave him 7 more.
Then he had 12 caterpillars.
How many fish did he have at the beginning?

Join – start unknown (2/11, 6/18, 4/11)
Mathtastic Level 2 – Numbers to 20
Add and Subtract by Compensation
tracyashbridge.com

Sam had 14 balls.
He gave 5 to Sarah.
How many balls does Sam have left?

Separate – result unknown (17/4, 12/3, 16/3)
Mathtastic Level 2 – Numbers to 20
Add and Subtract by Compensation
tracyashbridge.com

Dave had 14 cards.
He gave some to Tracy.
Dave has 5 cards left?
How many did he give to Tracy?

Separate - change unknown (13/6, 13/4, 18/13)
Mathtastic Level 2 - Numbers to 20
Add and Subtract by Compensation
tracyashbridge.com

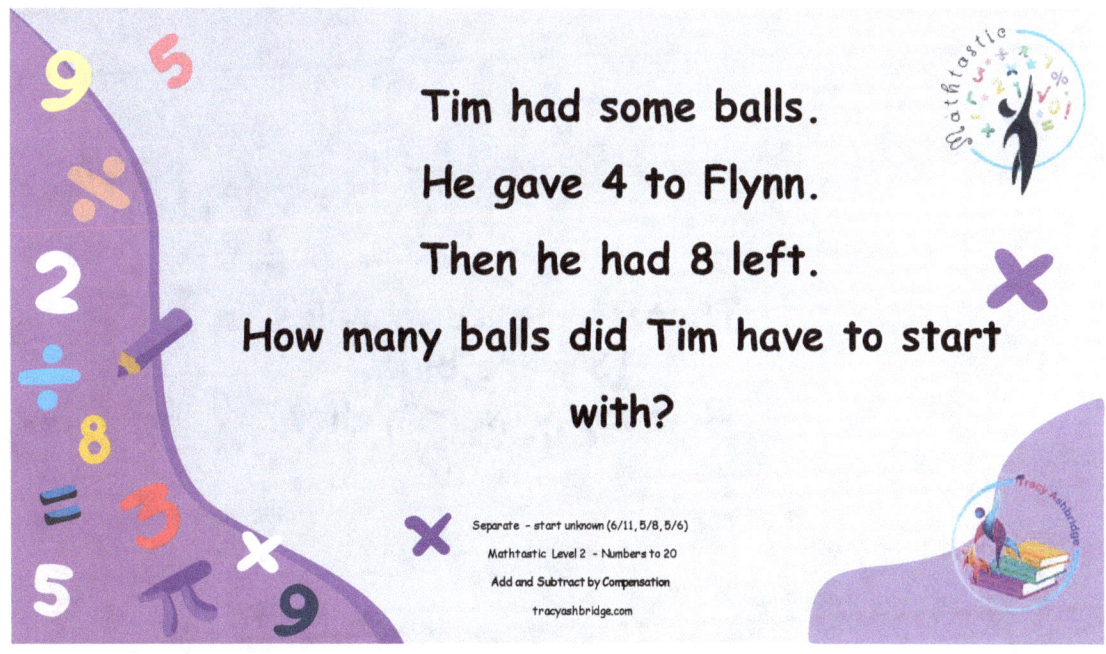

Tim had some balls.
He gave 4 to Flynn.
Then he had 8 left.
How many balls did Tim have to start with?

Separate - start unknown (6/11, 5/8, 5/6)
Mathtastic Level 2 - Numbers to 20
Add and Subtract by Compensation
tracyashbridge.com

9 yellow crayons and 6 red crayons were in the pot.
How many were in the pot altogether?

Part – Part– Whole – whole unknown (9/3, 8/4, 6/8)
Mathtastic Level 2 – Numbers to 20
Add and Subtract by Compensation
tracyashbridge.com

There were 18 marbles.
13 were blue.
How many were yellow?

Part – Part– Whole – part unknown (13/5, 16/9, 11/8)
Mathtastic Level 2 – Numbers to 20
Add and Subtract by Compensation
tracyashbridge.com

Georgie had 12 pens.
Jane has 5 pens.
Georgie has how many more pens than Jane?

Compare - difference unknown (13/7, 15/9, 12/8)
Mathtastic Level 2 - Numbers to 20
Add and Subtract by Compensation
tracyashbridge.com

Jason has 7 crabs.

Sam has 4 more crabs than Jason.

How many crabs does Sam have?

Compare - compared set unknown (8/5, 9/3, 12/6)
Mathtastic Level 2 - Numbers to 20
Add and Subtract by Compensation
tracyashbridge.com

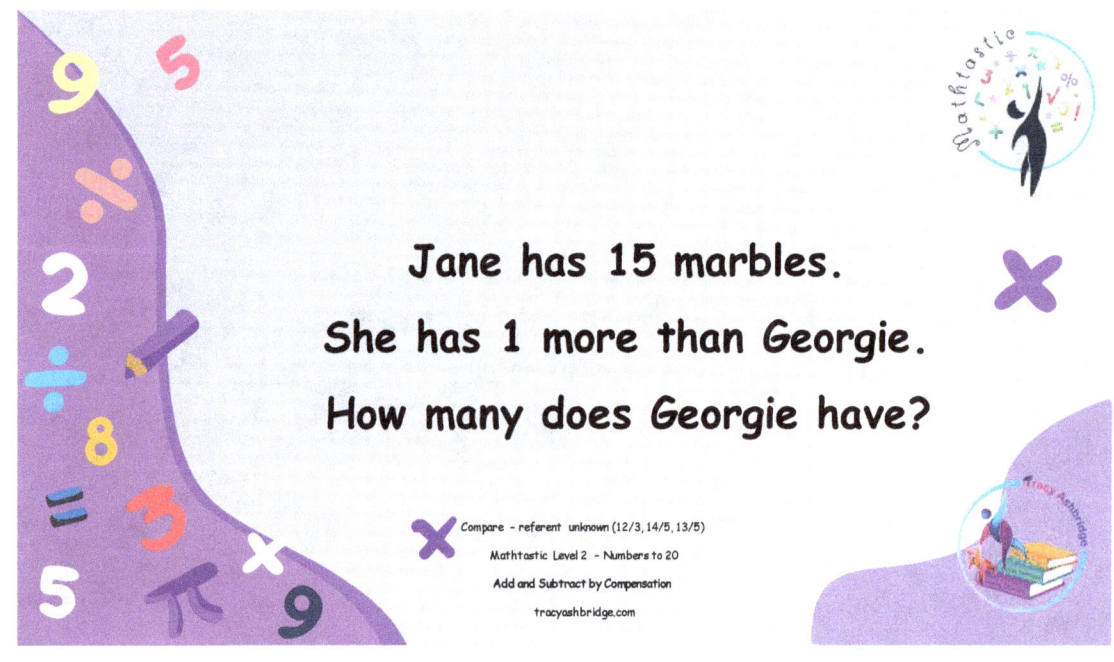

www.ingramcontent.com/pod-product-compliance
Lightning Source LLC
Chambersburg PA
CBHW080855010526
44107CB00057B/2586